# MANSFIELD PLANTATION

Christopher C. Boyle

*Foreword by John Rutledge Parker*

# MANSFIELD PLANTATION

*A Legacy on the Black River*

THE History PRESS

Published by The History Press
Charleston, SC 29403
www.historypress.net



First published 2014
Paperback edition 2015

Manufactured in the United States

ISBN 978.1.46711.774.6

Library of Congress CIP data applied for.

*For Brandon Joseph, Hannah Grace and the loving memory of Benjamin Charles Boyle.*
*With the love of my children, all things are possible.*

*I further dedicate this work to all of those who lived, dreamed, labored, loved, won and lost at Mansfield Plantation.*

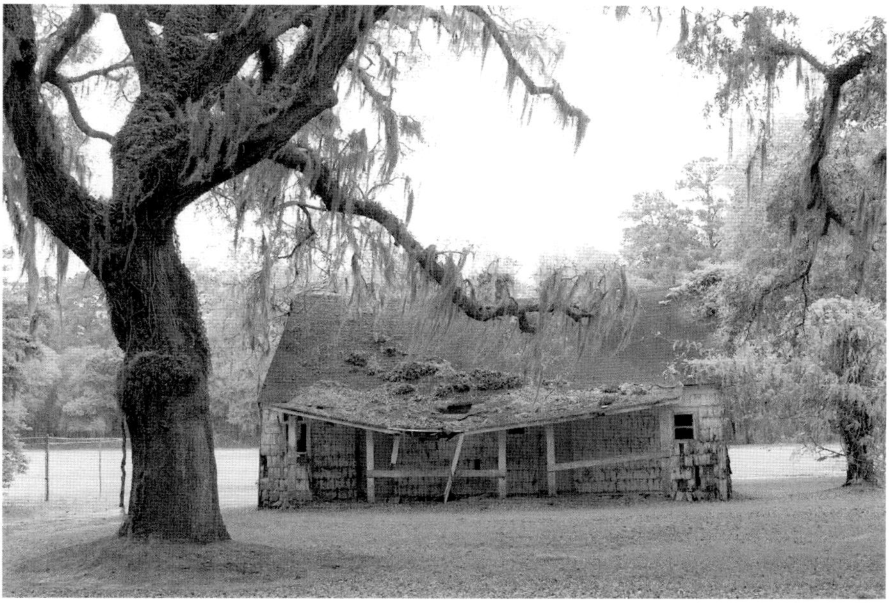

All author profits from the sale of this book have been lovingly donated to the "Resurrecting A Village" campaign. Two cabins and the chapel on Mansfield's street have been restored thus far, but the rest of the structures are in desperate need of repair. Your purchase of this book and any other donations are greatly appreciated. Please make your contributions to: The Dr. Francis S. and Mary L. Parker Foundation, 1776 Mansfield Road, Georgetown, SC 29440.

# Contents

# Foreword

From deep inside my earliest memories, I recall the name "Mansfield." For 156 years, it had been our ancestral home, located near Georgetown, South Carolina, on the banks of the Black River, and it was always spoken of in hushed, almost reverent, tones. It was sold in 1912, when my great-grandfather died at his summer residence in Flat Rock, North Carolina. The reason Mansfield was sold at his death was never discussed within our family.

Even as a small child, stories of Mansfield always held my interest, and I loved to sit beside my nurse, Becky (Rebecca Golden), and listen to her stories about her early life there and the little things she could remember that would bring a wide smile to her beautiful face.

Becky was born at Mansfield around 1900. She was the granddaughter of slaves and lived in their cabin until she moved with my family to Lake City, South Carolina. In essence, she was a part of our family. She died in the early months of 1968, and I lost my first and last direct link to Mansfield's past. My last vision of Becky was her being wrapped in a blanket, reclining in her chair and smiling at me. On the wall behind her were four pictures, each having a special meaning in her life: Dr. Martin Luther King Jr., President John F. Kennedy, my brother Captain Arthur M. Parker III (who would give his life in Vietnam six months later) and me. To this day, I remain humbled to be held in such esteem.

Becky would often use the saying "Never look back," which she had heard my grandfather Arthur Middleton Parker say to her as a child. Those words were important to her, and she made certain I knew and understood them. I have often wondered if he gave those words to her in order to somehow help her transition the difficult days they all endured. She made sure this simple wisdom was passed on.

She had an affinity for my great-grandfather that perhaps stemmed from her childhood memories of him returning from town at pretty much the same time every day. The children of the village would be waiting at Mansfield's entrance gate, and as he approached, driving his horse and buggy, they would swing it open for him to enter. They knew full well that as he passed, he would reach beneath the seat into a small brown paper bag and fill the air with candies as their earned reward.

Another of her favorite stories was how that same great-grandfather (who volunteered to join the Confederacy just after his sixteenth birthday) in one afternoon had two different horses shot out from under him at the Battle of Secessionville on James Island and lived to fight another day. I still have his Confederate Colt Navy .36 pistol, which I used many times as a child to reenact the great adventures that it surely had seen.

And then there was the hunting horn, which I was never allowed to touch as a child because it held such special meaning to earlier generations, as it was a gift from General Wade Hampton to my great-great-grandfather Dr. Francis S. Parker and represented a bond between the families. The horn was purported to be the general's personal hunting horn and was given to Dr. Parker on a hunting visit to Mansfield.

My first encounter of Mansfield was on a warm October afternoon in 1963 during a water skiing trip down the Black River. We noted a dock on the river that had a glass-paned enclosure with a small sign that read, "Mansfield." Feeling an air of teenaged excitement, we guided the ski boat down the long, narrow rice canal and finally reached a boat landing, where we started our self-guided tour. I was home from the moment I stepped from the boat onto the soft soil, and I vowed that if ever the opportunity presented itself, I would bring Mansfield back into the arms of the Parker family.

That opportunity—through a lot of prayer; love and support from my wife, Sallie Middleton Parker; encouragement from my best friend, Wyndham Meredith Manning; and financial assistance from my friends at First Citizens Bank—came about on April 28, 2004. My life was forever changed, as Mansfield was back. This is the story of her place in the rice culture of South Carolina. Thank you, Chris Boyle, for the research, writing and undying tenacity to make it happen.

*John Rutledge Parker*

# *Preface*
# Why Research Mansfield Now?

*Do not applaud me. It is not I who speaks to you, but history which speaks through my mouth.*
*—Fustel de Coulanges*

The tract of land referred to as "Mansfield Plantation" has seen constant occupation for the last two centuries by planters and their slaves, farmers and their employees and, beginning in the early twentieth century, caretakers. Before this relatively short period of time, the land hosted Native American settlements, and wildlife ran across her borders for millennia untold.

For over 150 years, planters and slaves worked the land and turned out tremendous yields of rice. After the American Civil War and the abolition of slavery, the Parker family attempted to hold on to the estate and continue rice farming. However, due to several conditions, rice-planting ventures collapsed.

Over the next century, northern investors purchased and sold the estate as a semitropical retreat from the harsh mid-Atlantic winters, a historic novelty and curiosity. First, the Tuttle family claimed Mansfield as their winter home, and then the Montgomery family took possession a generation later and converted it into a winter resort for hunting and social gatherings. The Montgomery family modified and modernized the estate to accommodate their needs and desires. The family also restored many parts of the estate and added electricity and a basement to the manor home, as well as to some of the other original structures. Under Montgomery's stewardship, Mansfield experienced a renaissance of sorts and became more ornate than ever before.

Within the previous two generations, proprietorship of the estate has returned to South Carolina natives. First, Wilbur Smith bought the estate from the Montgomery family in 1970. Later, his daughter Sally Smith took ownership. Among other projects, Sally turned the plantation into a bed-and-breakfast resort before eventually selling it to John and Sallie Middleton Parker.

John Parker, the great-great-grandson of Dr. Francis S. Parker, in coalition with his wife, Sallie Middleton Parker, established the Dr. Francis S. and Mary Lance Parker Foundation. A board of directors has been established to oversee various projects ranging from historical stabilization and preservation to the restoration of the structures on the estate.

The Parkers have fulfilled a dream that many southerners with similar family heritage have. It is altogether fitting that a little over 150 years after Dr. Francis S. Parker signed South Carolina's Ordinance of Secession and took that fateful leap that began the American Civil War—or, as he and the members of his propagation saw it, the War for Southern Independence—and a little over 100 years after the estate was originally sold out of the family hands, the history of Mansfield Plantation is finally written, her secrets revealed and her history shared.

# Acknowledgements

Many people deserve credit for assisting in the production of this book. First of all, John and Sallie Middleton Parker have made the completion of this project possible through funding the publication and always being supportive of this work.

Sallie Middleton Parker provided indispensable editorial support to the entire original manuscript. Without Sallie's assistance, I simply could not have completed this project. Glenda Y. Sweet, professor emeritus of English at Coastal Carolina University, edited portions of the manuscript in various stages, and I am truly thankful for her help.

Dwight Parker, board member of Mansfield Plantation and descendant of Mansfield slaves, provided many African American primary source records used in this work. Other board members—including Mansfield's current owner, John Rutledge Parker, and his cousin John Rutledge Parker (affectionately referred to as "the elder")—provided family stories and were especially helpful in assisting me to balance history and family lore to make this volume both personal and academic.

Dr. Kenneth Townsend, professor of history at Coastal Carolina University; Dr. Jason Silverman, professor of history at Winthrop University; Dr. Lynn Willoughby, professor of history at Winthrop University; Dr. Charles Joyner, emeritus professor of history at Coastal Carolina University; Dr. George C. Rogers, emeritus professor of history at the University of South Carolina; and Professor James L. Michie, associate director of the Waccamaw Center for Cultural and Historical Studies at Coastal Carolina

University and associate principal investigator and program director of the South Carolina Institute of Archaeology and Anthropology, University of South Carolina, all guided me with research and sources and questioned my findings in their own individual ways over the long course of my research.

My close personal friend Benjamin Burroughs, director of the Horry County Archives Center at Coastal Carolina University, provided me with an outlet for very serious historical discussions and helped to keep me focused on the project by challenging my knowledge of local history and pushing me to dig deeper into primary sources.

Another trusted friend, Rich Taylor, spent a few days at Mansfield and photographed the plantation for this project during different times of the day to get the best photographs possible. He also worked with archival photographs of the estate and helped me to select the best shots to include in this work. John and Sallie Middleton Parker provided the archival photographs used in this book, which they received as a gift from the Montgomery family.

Jim Michie and Susan McMillan, along with their crews of associates and volunteers, each led archaeological excavations that helped develop valuable insight into the lifestyles and folkways of Mansfield's residents. Hal McGirt also added to Mansfield's archaeological record with his own systematic method of recovering metal artifacts with detectors. Collectively, these findings helped me to unravel the story of Mansfield by examining the archaeological remains left by the estate's former inhabitants and visitors.

In researching the history of Mansfield Plantation and that of her proprietors and laborers, I leaned on the help of many people at many libraries. Employees and volunteers at the following depositories helped by retrieving documents and published works: the South Carolina Historical Society, the Southern Historical Collection at Chapel Hill, the Duke University Library, the South Caroliniana and Thomas Cooper Libraries at the University of South Carolina, the Kimbel Library at Coastal Carolina University, the South Carolina State History and Archives, the Chapin Memorial Library and the Horry and Georgetown County Libraries.

I would also like to extend my appreciation to my extended family and close friends who tolerated listening to my stories throughout the years of my research and also helped divert my attention from the studies of Mansfield when I needed breaks. My parents, Jerry and Annette Boyle, were especially instrumental in the development of this project, as they were always there for me and always instilled in me the belief that I can succeed in anything I attempt as long as my heart is in the project.

Lastly, a heart-felt thanks goes out to D'Andrea Dailey, a high school girlfriend whom I had not seen in twenty-six years until August 2013. An amazing beacon of light in my life, Dee found me emotionally broken and shell-shocked from my second divorce. She picked me up; dusted off my Mansfield manuscript, which had sat idle for over a year; and helped me to feel whole once again. After years of research and writing, I had given up on publishing this work—her love, understanding, trust, confidence in me and unselfish friendship served as the final push that brought this manuscript to print.

*Introduction*

# Mansfield Plantation: A Shrouded Past

*Bowered in its reticent grove of massive old live-oaks, there stands the ancient Southern Colonial home, beautiful, wistful, almost like some dreamy fair memory of days long gone.*[1]
—*Archibald Rutledge*

Throughout the South Carolina Lowcountry, signs of a past long ignored are in danger of being forgotten. Everywhere the old structures are slammed up against the new, the new quickly gaining momentum and the old withering away almost before our eyes. Urban sprawl has reshaped settlement patterns, and countless rice plantations have been consumed by an increasing population. The expanding city of Georgetown has swallowed the rice plantations of Waties Point, Greenwich, Willowbank, Richmond, Rosemont, Kensington, Weehaw and Windsor. The only reminders of their existence are the street signs that bear their names. Community developments and restaurants occupy grounds of the old plantations. Many of the elders who came of age working the fields of these estates, like the structures of the once mighty plantations, have passed on to their heavenly reward. Few have recorded their reminiscences, leaving modern man generations away from the truth about our plantation past.

Our plantation past is steeped in misunderstanding and contradiction. In some cases, decrepit structures have been bulldozed to make way for real estate development and golf courses. Some say they are making way for progress, while others welcome their disappearance from the Carolina coast in the hope that their history will be forgotten. Records have been revised or lost completely. Many are ashamed to seek the answers to questions that could help us gain a more accurate understanding of what transpired on these now almost fabled settlements.

Conceived here in the early 1700s at the twilight of Native American independence, the plantation system took root in the swamps of the Carolina Lowcountry, where the ebb and flow of the Atlantic tides wash the land with fresh water from the state's interior and provide a clean environment for rice to grow.

Today, superhighways crisscross the landscape. Modern man zooms past the old estates and over bridges where ferrymen once labored (until the 1930s) to shopping malls and community developments that use the title "plantation" so frequently and freely that most people hardly think twice of the implications of the word.

Progress ends at Mansfield's gates. The plantation is positioned two miles down a private dirt road on the tranquil banks of the Black River. Her history lies dormant, not completely forgotten but overlooked, not entirely beyond recall but distorted by revision and modern politics.

Nearly one thousand acres on the Black River of Georgetown, South Carolina, lie silent, peaceful and untouched by the ever-growing tide of land development. Once a self-sufficient community on the hither edge of civilization and home to over one hundred souls, only the nucleus of the great estate now stands. The wreckage of the mid-nineteenth-century property (the old steam-powered rice-threshing mill, the slave chapel and village, the winnowing house, the manor house, the kitchen and the school) remains to represent the once bustling economic domain. These structures stand today as reminders of our incomplete understanding of our plantation past: destroyed by war and neglect and finally abandoned as relics of an ancient people forsaken for their part in a failed revolution.

Among the casualties of abandon, the slave commissary, overseer's house, dairy, rice barns, icehouse and the entire eighteenth-century settlement pattern have fallen to the ground, some of their remains swallowed by the earth and forgotten. Others were deemed eyesores in disrepair and removed before they could be interpreted. They are now lost forever. Some of the more resilient structures remain, but in design and operation, some of these buildings, such as the winnowing house, are so simple that they are easily overlooked—their form and function simplistically complicated to modern man.

Mansfield bore witness to many of our nation's most colorful historical events and episodes. She observed beauty flourish and war destroy. Here, Susannah Man worked the land without the use of an overseer or the company of a husband. One of her sons-in-law, James Cassels, an officer in the King's Army, unsuccessfully defended the land from the American Revolutionaries for the king of England, while her future son-in-law Archibald Taylor became a leader of the Patriot cause. Upon these grounds, Susannah Man's grandson John Man Taylor, equipped with his Harvard degree and his appetite for social intercourse, expanded the estate to include over 125 laborers.

Within these acres, the Episcopal reverend Maurice Harvey Lance planted rice while he spread the words of the Christian Bible at Prince George Winyah Episcopal Church. It was also here that Dr. Francis S. Parker, intrigued by Mansfield and profits of rice planting, decided not to practice medicine professionally and instead joined the planter class. Here, Parker and his wife, Mary, planted the live oaks with their own hands and contemplated independence from the United States when South Carolinians felt their rights were being trampled on by the federal government. Meanwhile, at Mary's sister Ester Jane's Belle Rive Plantation, she and her husband, John Harleston Read II, argued against secession and remained loyal to the United States until the state officially withdrew from the Union.

After the upheaval of war and emancipation, the rice culture quickly declined, and Mansfield and her sister estates fell into disrepair and were sold out of the antebellum family.

They're all gone now, the continuum of the Man-Taylor-Lance-Parker original plantation family, and the Tuttle and the Montgomery families, and the Smith and the Cahalan families who followed them. The estate is now a bed-and-breakfast resort staffed by resident caretakers. The words "Tory" and "Loyalist," "secession," "tariff" and "Southern independence," "overseer," "master" and "slave" are no longer spoken in the present tense. Now, only visitors, relic hunters and trespassers enter the domain to reflect on the past.

For over two centuries, slave children and their descendants played games around the cook fires on the street while their parents mingled and socialized with generations of family members and friends recounting life's lessons and friendly gossip. Years ago, the chapel bell rang in church service on the estate, and the bullwhip encouraged bondsmen and laborers to keep up the quick pace necessary at harvest. The smell of boiling pork and game once filled the air around the servants' quarters. The Negro spirituals and work songs once lilted throughout the land, but now all of the voices are silenced. The sounds of the working community are gone, for the former fields of gold lie fallow and fruitless. Only the sounds of wildlife and the wind whistling through the trees can be heard.

Now the air is clear, except for the earthy fragrance of the Black River and the surrounding swamps. The manor house's kitchen, no longer staffed with domestic servants, does not produce fine meals. Floodgates, symbolic of the entire agricultural estate, worn by weather and left in ill repair, have slipped into the abandoned rice fields and no longer manipulate the flow of water into and out of the former fields.

The people who lived, dreamed, labored, loved, won and lost here are forgotten by time—the consequence of supporting a failed revolution. The proprietor Dr. Francis S. Parker, who played an instrumental role in the war, having signed South Carolina's Ordinance of Secession and maintained order in Georgetown as the provost marshal, is not remembered favorably or even with due respect in the histories composed by the victors of the war. Consequently, the significance of Mansfield has slipped past the compass of historians, through the hands of time and the annals of history.

Cattails now dominate most of the fields of gold where rice once grew. Aided by water tupelo and the occasional cypress, nature continues to reclaim these vacant fields. The uplands, a peninsula surrounded by the old rice fields and dense tupelo swamp to the southeast, contain the plantation nucleus where the planter, his family and servants once lived. This neck of land contains varied forests of pines, with sub canopies of emerging hardwoods, mostly oaks with some hickory, sweet gum, magnolia and American holly. Various other environmentally related species, including yaupon holly, wax myrtles and red bay, add to the developing pattern of plant succession moving toward the typical southern mixed hardwood forest. Upland fields, now fallow, once supported the working community with corn, peas, beans and grain but now are empty, recovering fields awaiting nourishment.

Here, masked in all of her mystery, all of her elegant simplicity, mighty Mansfield stands waiting to be rediscovered and analyzed by a new generation of objective interpreters. Her moss-draped avenue of live oaks houses countless pieces of pottery, remnants of her past. Each piece holds a different story, a clue to the secrets in her soil. The ground gives up more and more of her past with each rain, but Mansfield's history remains shrouded. Her future seems destined to follow that of other disintegrating plantations and become a resort or a housing development. What can possibly save her from this fate?

Enter John Rutledge and Sallie Middleton Parker, who purchased Mansfield on April 28, 2004. The Parkers initially started their renovations of the estate by restoring the manor house, including adding a new roof, painting the exterior and interior and rewiring the home.

The Parkers realized the entire slave village would soon be lost to the ravages of nature if something were not done. So they formed the Dr. Francis S. and Mary L. Parker Foundation, whose sole purpose was to do just that. The Parker Foundation consists of eighteen acres in the heart of Mansfield Plantation that contain the slave village and cemetery. Its board is made up of a diverse group of people, a few connected to Mansfield by blood and a few not. The thing they all have in common is a love of Mansfield and the realization that the slave village and cemetery are not to be lost.

Under the guise of the board of directors, various projects ranging from stabilizing and rebuilding the remaining cabins in the slave settlement and its chapel to the convergence of one slave cabin into a plantation museum have been conducted. The board has dedicated itself to maintaining and beautifying the plantation cemetery and has placed a marker to honor all those who lived, dreamed, labored, loved, won and lost and are interred at Mansfield. The foundation has also supported the second archaeological excavation of Mansfield, conducted by Susan McMillan and the Waccamaw Diggers, and the methodological metal detection work performed by Hal McGirt, as well as the publication of this history of the estate.

Beyond the work of the board, on December 26, 2013, John and Sallie entered into an agreement with the Nature Conservancy to place a conservation easement on 820.41 acres of Mansfield Plantation. This historical landmark will be protected from subdivision or development of any kind for perpetuity and remain much as it is for future generations to enjoy.

John Parker stated that it was his intent from the date that he and Sallie purchased Mansfield—April 28, 2004—to ensure Mansfield remained intact as a plantation unit and did not fall under the shovel of the developers who have subdivided many of the neighboring plantations.

*Chapter 1*

# Colonial Times: Susannah Man Carves a Plantation Out of the Wilderness

*These Carolinian gentlemen are well-bred, courteous and hospitable. A genuine aristocracy, they have time to cultivate their minds, to apply themselves to politics and the guidance of public affairs. They travel and read, love field sports, racing, shooting, hunting, and fishing, are bold horseman, and good shots. But after all, their state is a modern Sparta—an aristocracy resting on a helotry and with nothing else to rest upon.*

*—William Howard Russell*[2]

The Georgetown rice culture, though the most prosperous of the Lowcountry planting societies, was simply a microcosm of the greater culture of the American South. But to Georgetonians, it is believed to have held a special aura of lavishness unsurpassed in South Carolina or any other place in American history.

Politically and socially, Georgetown, South Carolina, like much of the antebellum South, was not a democracy but rather an oligarchy—a government and social structure controlled by the planter class. This small group of affluent individuals controlled all aspects of society in Georgetown and throughout the entire American South.

When reflecting on the rice culture, many people romanticize South Carolina's Lowcountry plantation past, while others demonize its planters and overseers. The purveyors of the romanticized version often visualize riding in a horse-drawn open carriage down a long avenue of moss-draped live oaks leading up to a palatial manor house. They celebrate the order and dominance provided by the planter class. The carriage passes laborers singing happily as they plant or harvest a field. Others, however, envision brutal punishments—an overseer's whip cracking down on a slave's bare back. The truths to the plantation past lie somewhere in the middle of these perceptions and are much more complex.

The plantation was an intricate, multifaceted unit. Its force was like a conscript army living in barracks. Husbands, wives and family members were comrades in service under the authority of a master and, to a lesser extent, a plantation manager or overseer. The master was captain in that he issued orders and also a quartermaster in that he issued food and clothing, whereas the overseer was a lieutenant in that he saw that orders were executed by the slaves, who acted like soldiers performing the work.

The plantation was a homestead and a school. Plantations were self-contained, isolated communities on the hither edge of civilization where food staples were produced for consumption and at least one product (in this case, rice) was created for export from that community. Just like in a school, ideas were shared among African slaves and European Americans. Slaves were trained to be field workers or skilled servants, and young masters learned how to run their plantations. The planter's children were nursed by female plantation slaves, and children of both master and slaves played together and learned from one another. On the plantation, too, master and slave adults worked together and enjoyed leisure time together, and from these interactions, cultural diffusion (the sharing of cultural traits and beliefs) took place. Generations of family members lived in the slave community and shared life's lessons, superstitions, wives' tales and unrecorded history.

The plantation was a factory with a mill village. As a factory, the planters provided compensation such as basic foods (mostly grown on the plantation), clothing and homes. On larger plantations, they provided central kitchens, day nurseries, infirmaries, visiting physicians and retirement for the aged. Planters were encouraged to keep their young slaves healthy to ensure that they reached adulthood. Upon the backs of the workers was the economic well-being of the community. As such, the morale of the workers had to be kept as high as possible to ensure the production of a profitable and marketable crop.

At the same time, the plantation was a parish where Christianity was used to teach slaves discipline and, of course, to help them to attain atonement for their lives of servitude. Some planters had churches constructed on their plantations for these purposes. With Christianity came sacraments such as marriage, which bound the slaves to one another and kept them committed to their mates, as well as to their masters and overseers by learning discipline and devotion as God's will.

Planters were compelled by the notion of deference and felt that it was their *noblesse oblige* to mandate politics and social order to their inferiors. Through their tremendous affluence, they created a world that inspired boys to become country gentleman and girls to become ladies. Children learned the art of manners and grace, and ladies married young and learned to obey their husbands. Collectively, the planter class held a great responsibility to maintain social order and provide for its slaves and for its less fortunate white neighbors as well.

During the era in which the rice planters ruled the Lowcountry, a distinct caste system prevailed. The system they embraced, one that was handed down from their progenitors in Europe, pertained not only to economics but also to race relations. This caste system, which held the planter class above the merchant class, poor white farmers, free blacks and, finally, slaves, eventually proved to be the pretext that led to the demise of their culture. In many respects, the fall of the planter class represents the end of a chapter in western civilization, when agriculture controlled the economy and large agriculturists governed the masses in a paternalistic fashion.

The Lowcountry rice culture, as well as the plantation system throughout the American South, is usually explained as a transplant, or continuance, of medieval European manorialism. The addition of race as a determiner of slavery made the system uniquely American. However, viewing the Lowcountry rice planters as transplanted lords of the manor discounts the fact that they competed economically in a capitalistic world with an internationally acclaimed product within global markets with worldwide financing.

How this hybrid economic capitalistic and manorial social doctrine took place is not a difficult question to answer; however, how rice initially got to America is a conundrum over which historians have argued for centuries.

One theory is that one of King Charles II's eight Carolina proprietors, the Duke of Albemarle, originally set up the colony of

Carolina to produce rice. In a letter dated August 31, 1663, the Duke of Albemarle wrote to Lord Willoughby, the governor of Barbados, discussing the future of Carolina, which, according to Albemarle, was ideal for the commodities of "wine, oyle, reasons, currents, rice, [and] silke." In the letter, Albemarle suggested that some of the Barbados planters move to Carolina to set up the production of rice and thus expand the Barbados plantation model.[3]

Another theory claims that Captain Thurber, a sea captain docked in Charles Town during the 1680s, gave Dr. Henry Woodward, an explorer who was left behind during the 1666 English expedition to Carolina, a bag of Madagascar rice. According to the story, Woodward, who knew that rice was in demand throughout the British Empire, planted his sample in the swamps near Charles Town. Supposedly, his experiment was a success, and the annual production of rice began.[4]

Yet another theory credits West African bondsmen with the introduction of rice and the knowledge to cultivate the crop in Carolina. This thesis, first developed by Peter Wood in 1974 and furthered by other authors such as Daniel Littlefield and Amelia Wallace Vernon, has become an interesting argument. Wood, in his book *Black Majority*, claims that Africans first grew rice in Carolina to subsidize their diets. Wood said that after the colonists witnessed the success of the crop, English colonists in Carolina, whom Wood asserts knew virtually nothing about the cultivation of rice, made a concerted effort to obtain bondsmen from rice-growing areas of West Africa.

Littlefield furthered Wood's research by studying the similarities between the cultivation of rice in colonial South Carolina and West Africa. In his book *Rice and Slaves*, Littlefield points to the survival of the African flail sticks for threshing the staple, fanner baskets for removing the tailings, the mortar and pestle for pounding rice and the African system of irrigating rice fields as evidence of transplanting the West African rice culture to the Lowcountry.[5]

By 1690, growers were producing rice in the South Carolina Lowcountry in large quantities. Colonists, with a shortage of currency and dissatisfied with Governor James Colleton's tax laws, petitioned that they be permitted to pay their taxes in commodities. On September 26, 1691, the General Assembly of South Carolina ratified the proposal in an act that allowed Carolinians to pay their taxes in rice as well as other goods.[6]

Once rice proved to be a reliable cash crop, prospective planters began to enlist a large labor force to improve the land. In the swamp areas around Charles Town, the main chore involved removing the cypress and gum trees and ditching and diking the land. Without a large labor force of white settlers present to perform these duties, and with Native American slaves continuously running away, South Carolina planters looked to Virginia's example of using Africans to perform these tasks.[7]

During its experimental stage, rice in Carolina was grown primarily in river swamps around Charles Town. In 1700, 330 tons of rice were exported from Charles Town to England and the West Indies.[8] Edward Randolph, collector of customs for the Southern Department of North America, wrote a letter to his superiors at the board, explaining, "They have now found out the true way of raising and husking Rice. There has been above 300 Tons shipped this year to England besides about 30 Tons more to the Islands." During the spring of 1700, the governor of Carolina wrote a letter to the Lords Commissioners of Trade and Plantations explaining the problems that Carolina was experiencing due to the overproduction of rice. He wrote, "[We] hath made more rice ye Last Cropp then we have Ships to Transport."[9]

In 1700, the George Town District was part of the Carolina frontier and was economically dependent on forest industries and the Native American fur trade. However, due to population growth and the success of rice and other local crops, land grants became available in 1705 for the tidal lands around the Waccamaw, Black, Sampit and Pee Dee Rivers.

It was during this time of expansion into the frontier north, on August 18, 1710, that John Green began to petition for a land grant of five hundred acres in what was then known as Craven County.[10] In 1718, the colonial government granted Green five hundred acres for five shillings.[11] This was the original plantation tract that eventually became Mansfield. The area was a wilderness. An Indian trading post operated less than fifteen miles north of Green's property.[12] John Green held the deeds to other lands that later became the plantations of Windsor, Wedgefield, Beneventum and Peru/Cumbee. These lands were all laid out on a creek flowing into the Black River from the south that soon came to be called Green's Creek.[13]

During the 1720s and 1730s, the South Carolina rice industry experienced its first boom, but the new George Town planters placed more concern on building their plantations and attempting to find new ways to cultivate rice than on the production of the crop.

Beginning in 1726, in attempts to promote settlement, the colony passed acts for improving existing roads and building new ones, as well as building bridges over the many rivers in the region since ferries were very slow and unpredictable.[14] George Town was laid out in 1729, and in 1732, it became an official port of entry.[15] By that time, most of the Native Americans in the district had been destroyed by war, disease or enslavement. Those who had survived had been pushed back from the George Town District and farther into the Carolina frontier, where they either joined other tribes or perished.[16]

At the onset of the War of Austrian Succession between England and Spain in 1740 (known in the colonies as King George's War), George Town exported 4,785 barrels of the staple crop, while Charles Town merchants exported 80,000 barrels. Due to crop failures between 1741 and 1746, as well as war, the George Town 1740 exportation figure was not matched again until 1755 and not substantially surpassed until 1761, when a similar pattern occurred in Charleston.[17]

In 1750, John Green died during the lull in rice profits and the period of building the George Town rice district. In his will, he instructed his wife, Elizabeth, to sell the property and divide the revenue among herself and their three children: John, Richard and Elizabeth.[18]

In March 1754, James Coachman (a land speculator who already owned two plantations on the Black River as well as other properties)[19] and his wife, Hannah of St. James in Goose Creek Parish, purchased the estate.[20] Coachman also bought the land adjacent to the property, which later became Beneventum Plantation.[21] Meanwhile, the production of rice was advanced to the flat areas near the creek bottoms, but it was not until the 1750s that the rice culture was expanded to the tidal lands.[22]

In 1754, a new English war threatened trade. By 1756, the Seven Years' War (known in the colonies as the French and Indian War and in the Carolinas specifically as the Cherokee War) spilled over from North America to include the European powers. When the French navy began to seize ships, it crippled colonial freight shipping, and the rice trade suffered. In 1763, the English forcibly removed the French from North America and restored colonial shipping. By 1766, the colonial rice trade had stabilized, and rice prices remained high until the outbreak of the American Revolution.[23]

Unlike the traditional view of the southern colonies and later states, during the colonial period, the people of George Town were committed to a diverse economy. The South Carolina ship register from 1734 to 1774 reveals that thirty-three ships were built in George Town. George Town was, behind Charles Town, the most important shipbuilding site in South Carolina until the eve of the American Revolution, when it was surpassed by Beaufort. The people of the pine forest provided the raw materials of tar, pitch and turpentine needed for shipbuilding.[24] George Town was situated on the Atlantic seaboard, and since the land was disrupted by so many rivers, the main route of transportation was by water, thus making shipbuilding a necessity.

Throughout the colonial period, Northern Europe (mainly Holland and the German states) imported as much as 65 percent of the rice exported from South Carolina and Georgia. Southern Europe (primarily Spain and Portugal) imported less rice than Northern Europe. Southern Europe imported only 17 percent of the Georgia and South Carolina crop.[25] Shipments to the West Indies accounted for the remaining 18 percent. Generally, English consumers eschewed rice and remained dependent on their staple of wheaten bread. France banned the importation of North American rice and, like the Spanish, relied primarily on cheaper rice from Turkey and Brazil.[26] Thus, the largest market for Carolina planters was Northern Europe. In those regions, people and livestock ate rice during the winter, when peas were scare and barley was unavailable. To the poor, rice was a cheap alternative to other cereal products and local foods. Rice also supplemented bread and even meat in the diet of the non-poor. The greatest demand for rice on the Iberian Peninsula occurred during Lent and in times of substandard wheat or barley harvests. Of all the Carolina rice markets, the consumers of the West Indies received the poorest-quality rice, as planters bought the inferior and cheaper rice for consumption of their slave laborers, who worked the very large and profitable sugar plantations.[27]

A major influx of new field hands to the Lowcountry came from the African rice-growing areas of Angola, the Gambia River area, the Windward Coast and the Gold Coast. Planters preferred Africans from these areas, although some bondsmen, mostly prisoners of inter-tribal warfare in Africa, came from as far as seven hundred miles into the interior.

In African coastal regions, rice growing had been a dominant part of that culture since 1500 BC.[28] The slave trade in West Africa had existed long before the Europeans first stepped foot on the continent. The most common reasons why Africans sold tribal members to Europeans were for offenses against society, such as murder or theft; offenses against the king; or even due to personal or tribal indebtedness or famine. Bondsmen were trained in planting rice in Africa before they were sold to the Europeans, thus enabling African kings to receive higher profits from the sale. Once sold to the New England and English shippers, the bondsmen were transported to the southern planters, who needed laborers to build and cultivate agricultural staples on their estates.

At this time, two-thirds of South Carolina's population was represented by Africans. The use of local Indians as a labor force had dwindled. Large-scale Native American forced labor in South Carolina ended during this period for two reasons: (1) the Indians proved to be far weaker than the Africans (many were worked to death, and others died from white diseases) and (2) they also proved to be nearly impossible to retrieve once they escaped from their masters. Once freed, they could easily find groups of Indians and return to their traditional lifestyle.[29] In an attempt to end these problems, planters began to deport and sell their Native American slave laborers to the northern colonies of Massachusetts, Connecticut and New York, as well as the West Indies.[30]

Between 1735 and 1740, 12,589 Africans were imported into Charles Town. The 1740s, a decade that had considerably low rice exports, saw a decrease in the influx of laborers as well. During that decade, only 1,562 Africans were imported. The decrease might have been the result of the 1739 slave uprising south of Charles Town known as the Stono Rebellion, the largest such rebellion in colonial American history. Over the next twenty-five years, the trade reached new heights of importation. By 1775, English and New England traders had brought nearly 60,000 more Africans to South Carolina.[31]

In the early days of Carolina, from the 1690s to 1730, planters and slaves cultivated rice in upland impounded areas. With the development of "tidal flooding" in the 1730s and its perfection by 1750, George Town became the center of rice production in the colony since the area was drained by the tidal rivers: Waccamaw, Black, Pee Dee, Sampit and Santee.[32] Due to tidal flooding, rice became so profitable that merchants, physicians and attorneys left their professions, purchased land and slaves and became planters.

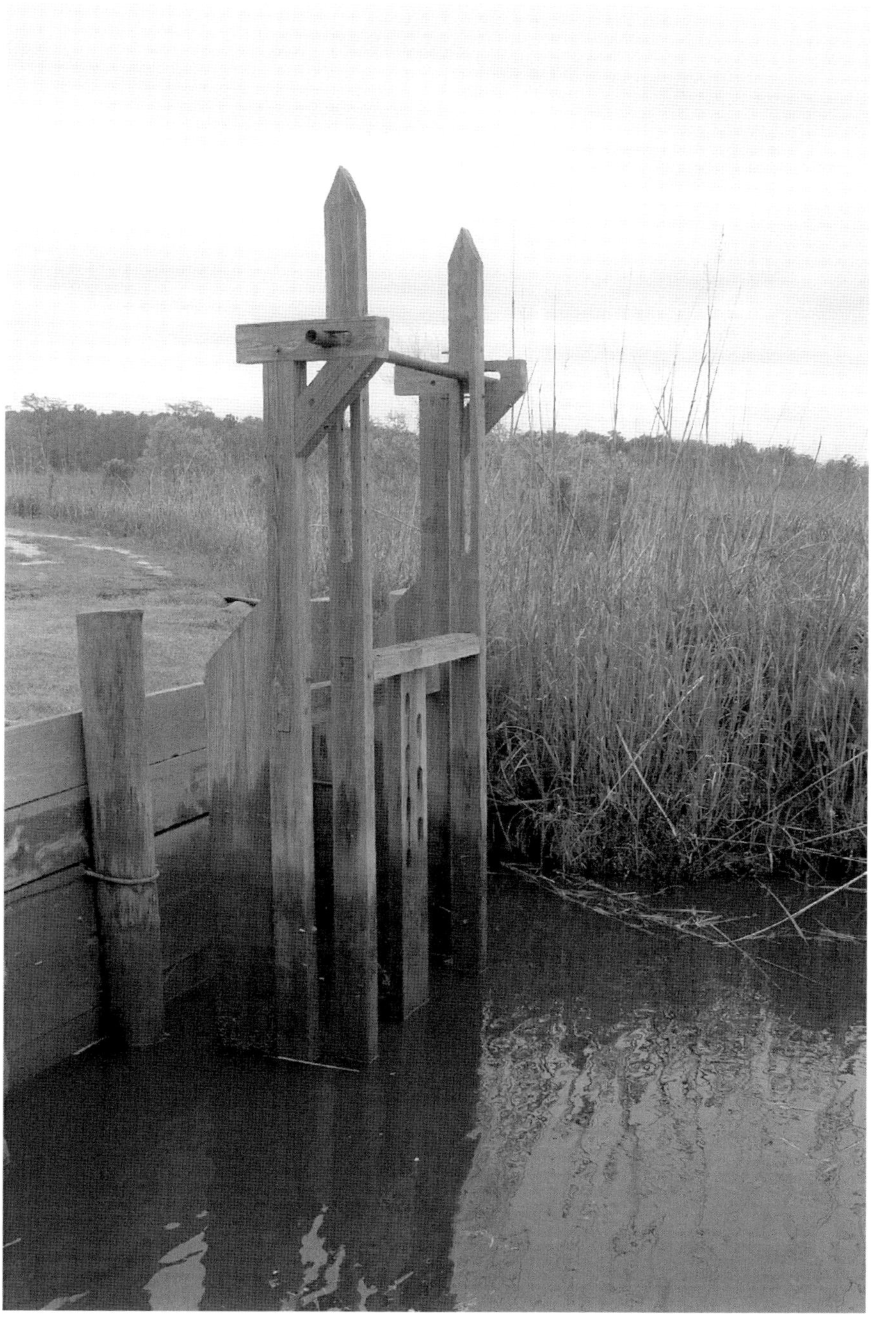

The floodgates controlled water in and out of the fields. Trunk-minders would taste-test the water for salt before they opened the gate to allow water in. The act was usually performed by a young boy.

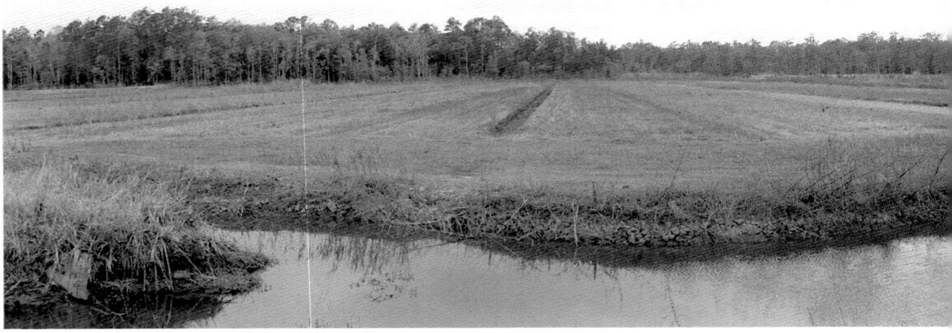

Although fallow now, these were once fields of plenty. The Parkers are planning to plant a sample crop in the spring of 2015.

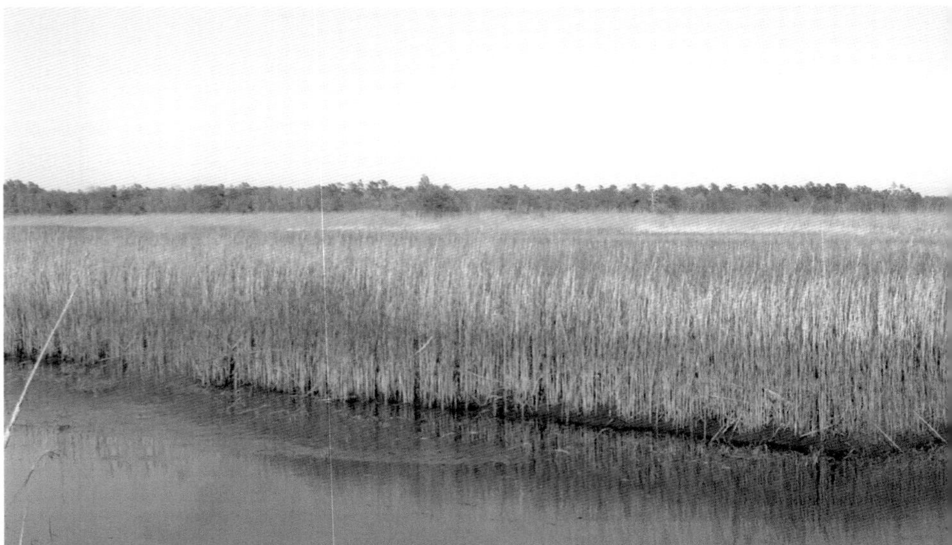

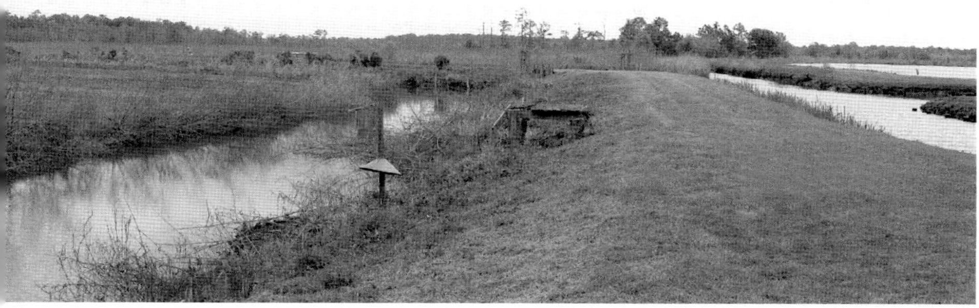

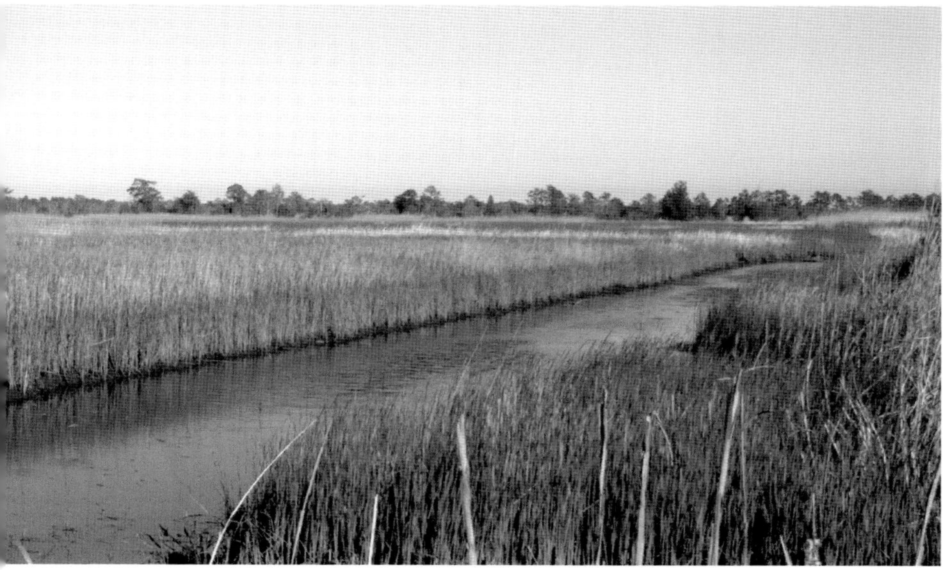

Fresh water was vital to rice farming. Tidal flooding allowed planters to grow rice along rivers anywhere river water could be reliably trapped by floodgates, contained and drained when necessary. At high tide, floodgates were opened to allow river water to fill the fields, which were then drained by opening the floodgates at low tide. The practice required several feet in pitch of tide (the difference between high and low tide). This system allowed plantations to be constructed farther into the interior of South Carolina on rivers with ample watersheds. Planters had to be careful to construct rice fields far enough away from the ocean because salt water would kill their plants.

The process of transforming the Carolina swamps into tidal rice fields took years to complete. First, the land had to be measured and trees felled and burned. Next, slaves cut a four- or five-foot ditch through the clearing. The mud cleared was used to form an enclosing bank, leaving a fifteen- or twenty-foot-wide space surrounded by a canal. When slaves completed this task, they put trunks into place and drained the land. After draining the fields, trees were felled, their stumps removed and together burned. Finally, the canals were temporarily blocked off, and floodgates were constructed. After digging the canals and building the floodgates, slaves planted the first crop in the new field.

The following season, laborers dug smaller ditches, referred to as quarter ditches because they were a quarter of an acre apart, running the length of the field. They repaired banks so that they remained a few feet higher than the spring tide and constructed smaller banks to divide the fields. Finally, bondsmen dug large ditches around the entire field, and small drains were put in place so that the fields could be drained.[33]

With the technology of tidal flooding, rice became the region's number-two cash crop; indigo was still number one. During the 1740s, the War of Austrian Succession had threatened the indigo crop in the West Indies and enabled the South Carolinians to gain the opportunity to grow great crops of indigo for the British. Prosperity for the South Carolina indigo market continued throughout the 1750s and 1760s due to the Seven Years' War.

The money generated from the production of indigo enabled the citizens of the community in 1757 to develop the Winyah Indigo Society, a philanthropic organization that had the honor of being the first societal order formed by the planters of George Town. It was originally formed as a social organization to share the news from Europe and the other English colonies but later became a strong link between the interests of the merchants and planters. During the American Revolution, however, the indigo market was hurt due the closure of commerce with the British Empire and never fully recovered.[34] By 1800, without a bounty (a farm subsidy) provided by the king, the production of commercial indigo in Georgetown had ceased. The legacy left by the crop is the continuance of the Winyah Indigo Society.

By 1750, the development of tidal flooding had made rice planting in George Town very profitable, and planters were scrambling to purchase swampland. In 1756, in this land-frenzy environment driven by the rice boom, James Coachman sold several tracts of land, including what would become Mansfield, to Mrs. Susannah Man, widow of John Man, the Prince Frederick Parish planter, physician and land speculator.[35]

Susannah Man, unlike the previous owners of this land, who simply bought it and held it for its improved resale value, moved to the estate. She constructed a house on the property and began the laborious task of transforming the Carolina swamps into gardens of golden rice. With her labor force intact and house and fields completed, Mrs. Man formally named the plantation Mansfield in memory of her husband, who died in 1754. Thus began a 150-year rice-planting tradition among her family members at Mansfield Plantation.[36]

Mrs. Man was the daughter of French Huguenots John and Mary Horry Laroche. She married Dr. John Man on April 7, 1743.[37] When her brother Thomas Laroche died later that same year, Susannah became a landowner as prescribed in his will. According to the document, she was to share ownership of the Bluff Plantation on the Pee Dee River, later renamed Birdfield, with her four sisters.

During her eleven-year marriage, Mrs. Susannah Man gave birth to two daughters. Susannah, the eldest, was born on April 6, 1745, and baptized on November 30 at Prince Frederick, Winyah. Two and a half years later, on May 12, 1748, Dr. and Susannah Man had Mary and baptized the child on May 31 at Prince Frederick, Winyah.[38]

Jointly, the Mans acquired what later became Springwood Plantation. At the Springwood site, however, Dr. Man got himself into considerable trouble. In 1750, Dr. Man's neighbor, John White, filed a bill of complaint against Dr. Man for trespassing and breaking into his home with force:

> *The plantation and the house and buildings of the said Anthony thereupon standing erected and built he did break and pull down; and the soil of the said Anthony he did subvert and dig up; and the wood, timber, and lightwood of the said Anthony thereon standing, growing, and being he did destroy, cut down, and consumed, and other enormities to the said Anthony to his great damage and against the Peace of our Lord the King.*[39]

Dr. Man pleaded not guilty to the charges, and history does not record how the case was resolved.

Dr. Man's estate reveals what he owned at the time of his death. Among his possession were brass candlesticks, a backgammon table, a parcel of sugar loaf, indigo seed on hand, rice sieves, a rice fan and forty-eight slaves.[10]

In 1769, Susannah, the Mans' elder daughter, married James Cassels, deputy tax collector and customs official in George Town and planter of Winyah and Charles Town. After giving birth to their son and after only one year of marriage, Susannah died. The infant son was left to her mother's care (Susannah Man) at Mansfield.[11] James Cassels went on to become a colonel in the British army and the leader of the Tories in the George Town area during the American Revolution. After the war, Cassels fled South Carolina via east Florida to England and died in Leith, Scotland, on March 20, 1798.[12]

In 1770, Mrs. Susannah Man purchased an additional tract of four hundred acres between the Black and Pee Dee Rivers to add to her rice empire.[13] Susannah also inherited from her mother eighty-one acres that later became the rice fields of Willowbank Plantation on the Black River.[14]

Mary Man continued to live with her mother at Mansfield until 1772, when she purchased lots numbered 41, 42 and 69 on Front and Queen Streets in George Town. The following year, Mrs. Man followed her daughter's lead and purchased adjoining lot number 43. At that point, Susannah and Mary officially moved to George Town.[15] Although six miles from the plantation, tradition has it that Mary rode out on horseback from George Town daily to oversee the production of rice and other crops at Mansfield.[16]

In February 1775, on the eve of the American Revolution, Mrs. Man sold sixty barrels of rice to the Wragg and Smith Company in George Town.[17]

During the American Revolution, the George Town planters became closely aligned to the Charles Town merchants and created closer social, political and economic ties with the people of the city. This shift toward dealing with the larger city market gave even greater credence to Charles Town becoming the center of plantation society as a new era dawned.

# *Chapter 2*

# Building a Family Tradition at Mansfield under John Man Taylor and Maurice Harvey Lance

*The ideal which took root in the South was distinctly that of English country life in which the owners of broad acres should also constitute the dominant class in government… The peaceful Southern planter played no such part as had the feudal lord of the past, but his large plantation, remote from others and on which he exercised a lavish hospitality, was nearer to the English country estate of the upper middle class than anything else in America…Families intermarried and built up cliques which ruled socially and politically. The aristocratic Southerner laid far more stress on family, land, agreeable manners, and political power than he did on mere wealth; and the long days on the plantation led him to cherish social life as an art.*
*—James Truslow Adams*[18]

After the American Revolution, American-grown indigo was no longer important to the British. The British simply restored their crop in the West Indies and created new markets throughout their empire. Planters turned their attention to rice and recorded greater earnings than ever.

The invention of the rice-pounding mill helped to foster its new position in the economy of South Carolina. Historians have often noted that the invention of the rice-threshing and pounding mills had comparable impact on the rice industry as Eli Whitney's cotton gin had on cotton production. With these new inventions, planters had no problem advancing the profitability of their crop.

Before the invention of the rice-pounding mill, slaves milled the grain with wooden hand-held tools called a mortar and pestle set. Milling, or pounding, as it was called, was a long and tedious process that took so long that some slaves were often still milling one crop while other laborers were preparing fields for the following year.

Animals, mostly oxen, provided the power to run the first mills. Then, in 1787, Jonathan Lucas built the first water-powered rice-pounding mill, and in 1792, Georgetown built its first tide-operated rice mill.[49] By this time, enough wealth was being generated by rice, and the price of milling was affordable enough, to allow planters to bring their "rough" rice to be prepared for market so they could give their slaves time to properly prepare the fields for the following year's crop.

Over the next seventy years, Georgetown continued to develop into a premier rice-producing region and eventually grew nearly half of all the rice produced in the United States. Soon, an aristocratic faction based on the production of the crop emerged. Cotton eventually ruled the southern economy in general, but the relatively small Georgetown District remained devoted solely to rice.

Since growing rice was such a labor-intensive trade, Georgetown needed a multitude of slaves to build its rice culture. The census report of 1790 shows slaves to be the majority of the population. In that year, the three parishes of Georgetown contained 22,122 persons, of whom 13,131 were slaves. At that time, the Prince George Winyah Parish, which contained the town of Georgetown, including Mansfield, had the most closely balanced population, with 5,111 whites and 6,651 slaves. Prince Frederick Parish, located northwest of Prince George, contained 3,450 whites and 4,685 slaves. All Saints Parish, a coastal community located east of Prince Frederick, consisted of 430 whites and 1,795 slaves.[50]

In 1791, the developing rice-planting aristocracy entertained President George Washington when he visited Georgetown on his first southern tour of the United States. President Washington wrote of his experiences in Georgetown in his diary:

> *Crossed the Waggamau* [sic] *to Georgetown by descending the River three miles at this place we were* [received] *under a salute of cannon and by a company of infantry handsomely uniformed. I dined with the citizens in public, and in the afternoon, was introduced to upwards of 50 ladies who had assembled at a tea party on the occasion.*[51]

Washington went on to write, "Georgetown seems to be in the shadow of Charleston. The inhabitants of this place (either unwilling or unable) could give no account of the number of souls in it, but I should not compute them at more that 5 or 600."[52]

Although the wealth of the planters who lived in the countryside continued to expand, the town of Georgetown grew very little. In 1826, when architect and surveyor Robert Mills visited Georgetown, the town's population had barely grown since Washington's visit thirty-five years earlier. Mills recorded, "The White population is between 6 and 700, and the Black about 12 or 1,400."[53] He also made note of the peculiar use of rice that he witnessed in the district: "In Georgetown everything is fed on rice; horse and cattle eat the straw and bran, fowls, etc. are sustained by the refuse; and man subsists upon the marrow of the grain."[54]

In 1785, Mary Man, at the age of thirty-seven, married Georgetown merchant Archibald Taylor and bore him two children: John Man (1786) and Anna Maria (1787). While a merchant, Mr. Taylor was also an investor in at least one rice plantation. In 1786, he became a partner in the firm Robert Collins and Company, which owned and operated Grove Plantation on the Black River. Taylor's partners in the firm were Robert Collins and John Mason, both Georgetown merchants. By the following year, Taylor had sold his interest in the firm.[55] In 1798, Taylor was chosen to be joint commissioner for clearing and making navigable Green's Creek in Georgetown.[56]

Even though she lived in town with her family, Mary Man Taylor continued to ride out to Mansfield daily to manage the estate. After the birth of her children, she summered in the North. On June 8, 1801, Mary died while on vacation in Boston, and her mother, Susannah Man, died two years later and left Mansfield and Willowbank Plantations, more than three hundred acres on the Sampit Creek and her entire estate to her seventeen-year-old grandson, John Man Taylor.[57] Archibald Taylor did not gain title to Mansfield Plantation or any of the other family holdings upon his wife's death because Susannah Man still legally owned the property. Archibald Taylor died in May 1821 and was buried on May 22, 1821, at seventy years old.[58]

John Man Taylor studied at Harvard, where he earned his bachelor's degree in 1803 and his master's degree in 1806.[59] In 1803, Taylor belonged to the fraternal order of Phi Beta Kappa and, upon graduating with his master's degree, joined the university's quasi-secretive society, the Porcellian Club.[60]

After graduation, John Man Taylor moved home, and unlike his mother and grandmother, he lived full time at Mansfield with his 125 slaves.[61] An effective manager of his labor force, Taylor often leased his slaves to other planters when work at Mansfield was completed, including his future brother-in-law Reverend Maurice Harvey Lance and his future niece's family, the Reads.[62]

Anna Maria Taylor married Josias Allston in 1804 at seventeen and became a widow by her twentieth birthday.[63] Showing his paternalistic impulse and expressing his family leadership role, John Man Taylor published an open letter in the *Georgetown Gazette* expressing his concerns for his young widowed sister's economic well-being. He wrote, "All persons indebted to the estate of Josias Allston, late of Georgetown, are requested to make immediate payment [to his widow]."[64]

The Lowcountry notion of *noblesse oblige* compelled Taylor to run for public office beginning in 1809, when he was elected as justice of the peace for Prince George Winyah Parish. The next year, the citizens of the parish elected him to represent them in the Nineteenth General Assembly from 1810 to 1812. Taylor was defeated in 1812, when he ran for a second term. Among Taylor's supporters for the House seat was the elder American Revolutionary War hero General Peter Horry.[65]

On April 18, 1812, Taylor increased the size of his gang by purchasing sixteen slaves from the estate of Thomas Shubrick.[66] By this time, Taylor was financially secure and interested in expanding his holdings in both land and slaves. Besides his inheritance of Mansfield and Willowbank Plantations (located on the lower Black River just north of the town of Georgetown), John Man Taylor soon purchased Smerden Plantation on the Pee Dee River from Priscilla Smerden after her husband, Elias, died in 1813.[67]

Like his father before him, Taylor concerned himself with social, financial and internal improvements. Taylor served as an incorporator of the Pee Dee Steamboat Company in 1819, director of the Georgetown branch of the Bank of the State of South Carolina from 1820 to 1822 and commissioner of roads for Prince George Winyah Parish in 1823.[68]

In addition to being a politician and planter, Taylor was an Episcopal vestryman at Prince George Winyah Church, a member of the Winyah Indigo Society and a member of the Georgetown Library from its inception in 1799 until his death.[69] A philanthropist and strong proponent of education, Taylor once donated one thousand books from his personal library to the Georgetown Library Society.[70]

Even while providing community services, gentlemen of the era brought their love for discourse and conviviality with them. In 1812,

General Peter Horry helped the Georgetown troops prepare defenses near the planter's popular summer resort on North Island in case the British should attack (the War of 1812 had already started). The fortification assignment allowed him to catch up with friends, including John Man Taylor. Horry noted in his diary that he visited Taylor at his summer refuge on North Island for breakfast and company until noon at least twice and dinner and alcoholic drinks on several occasions during the summers of 1812–14.[71]

The American Revolutionary War hero mentioned in his journal that Taylor's sister Allston (implying the widow of Josias Allston, Anna Maria Taylor) was among many members of the planter class who summered in the North. Anna Maria Taylor often passed the sickly season in Newport, Rhode Island, with her father. General Horry stated in his journal on September 8, 1812, that "she should be on her way home" and would soon join them at North Island.[72]

Being a coastal region, Georgetown's fortifications were a constant concern of the United States government even during peacetime throughout the eighteenth and nineteenth centuries. On April 24, 1819, the *Winyah Intelligencer* reported the visitation of President James Monroe to Georgetown. Secretary of war and South Carolina native John C. Calhoun and fellow South Carolinian Major General Thomas Pinckney, as well as others, accompanied the president. The president and his guests "examined Winyah Bay, the entrance to the port and the adjoining shores and waters."[73]

Planters preferred the resort at North Island, but other seaside resorts for summer leisure existed on Pawleys Island and at the marsh-front resort of Murrells Inlet. When engineer and architect Robert Mills conducted his tour of the state to record statistics of each district, he wrote favorably about North Island:

> *The inhabitants of Georgetown, and its vicinity, have a delightful and salubrious retreat in the sickly season, on North Island, and the adjacent Sea Islands. A happier situation is not to be found anywhere; for here perpetual breezes and saline vapors are constantly rising from the ocean. Three hours bring the citizens from the town to the sea. The good things of this life are here really enjoyed by the inhabitants in abundance; for the land and the ocean lay their treasures at their feet.*[74]

Later in his report on the Georgetown District, Mills recalled that the planters who summered at North Island constructed a church on the island. Surprisingly, he went on to say, "The number of persons which assemble on this island, at this season, is between 600 and 700." Considering that, according to Mills, Georgetown consisted of only 600 to 700 whites and about 1,200 African Americans, the resort was quite large.[75] On September 27–28, 1822, a hurricane hit the resort, and nearly 300 people (both slaves and masters) died in the storm.[76]

The 1820 census report shows that John Man Taylor owned 134 laborers, of whom 82 were engaged in agricultural production. A total of 33 members of his gang were listed as children under the age of fourteen, and 19 were listed as being over forty-five.[77] For his era, Taylor was a rather large slaveholder. The entire district consisted of 17,603 people, of whom 15,546 were slaves.[78]

In May 1821, during the Taylor occupancy of Mansfield, angry slaves murdered fellow Black River planter George R. Ford. The state called out the militia, and the murderers were tracked and finally caught by the local militia in the Santee swamps, but the incident proved that slave insurrections were possible at any time.[79]

The following year, one of the most fearful episodes in South Carolina history took place. Denmark Vesey, a free person of color living in Charleston, planned one of the largest slave rebellions in the United States. The plan was to set several fires in the city, at which time whites would respond to the fire, and the slaves would kill their masters as they rushed to their horses to fight the blaze. With the city left unprotected, freedmen and slaves alike would steal the money from the banks and take ships from Charleston to Santa Domingo. Word of the plans leaked, and Charleston authorities arrested the plot's leaders before the uprising could begin, which Vesey planned for Bastille Day, July 14, 1822. The South Carolina courts tried Vesey and others, ultimately convicting and executing them.[80]

Another failed slave insurrection took place in Georgetown in late July 1829. After rumors circulated about a possible rebellion, two slaves were apprehended, tried and hanged for attempting to lead an uprising. Wood, a slave of Black River planter Francis Kinloch, and Charles, a slave of Pee Dee River planter John Coachman, were found guilty by a court of justices and freeholders. John Coachman freed a slave named Henrietta and her son after she proved to be a reliable informant in the case of the planned uprising. Coachman and

Kinloch filed petitions with the magistrate and freeholders court for compensation for the slaves.[81]

In response to the failed insurrections and violence, South Carolina took extreme measures to ensure that another such incident did not take place and that, if it did, the state would be prepared. In 1690, long before Denmark Vesey's failed insurrection, South Carolina had already enacted an "Act for the Better Ordering of Slaves," and the colony passed even stricter slave codes after the Stono Rebellion of 1739, including limiting their access to English law, reading and writing; forbidding the playing of drums and horns; and even placing regulations on clothing. In addition to these older laws, new laws provoked by the Stono Rebellion included acts against slaves owning firearms and the hiring out of their own services. The new laws also included the Negro Seaman Act, which required that free blacks remain in quarantine when they entered ports. Over the next twenty years, the South Carolina General Assembly outlawed manumission as a measure to cut down on the number of free persons of color in the state.[82] Next, the South Carolina legislature passed an "Act to Convert the Arsenal at Columbia, and the Citadel and Magazine in and Near Charleston into Military Schools," thereby transforming the two state arsenals into the South Carolina Military Academy. The act specified that "the students, when admitted, shall be formed into a military corps, and shall constitute the public guard of the Arsenal at Columbia, and of the Citadel and Magazine in and near Charleston…to guard effectually the public arms and other property at the places aforesaid."[83]

A year after the failed Denmark Vesey insurrection, John Man Taylor went on a tour of England to visit his kinsman and only first cousin, John Cassels. John (born in 1770) was the son of James Cassels and Susannah Man. During the American Revolution, James Cassels served as a captain in the king's army. He was banished after the war and his plantation confiscated. Shortly after the war, in 1788, Archibald Taylor, Mary Man Taylor and Robert Heriot petitioned the state legislature to allow John to keep the family slaves. The case ended in John's favor, and his property rights were secured. After earning his law degree, John moved to London, where he died in 1830.[84]

When John Man Taylor returned home from England, he contracted what the people of Georgetown then called "country fever," which proved terminal. Even in his final month of life, Taylor remained devoted to his principles and fought for what he believed was best for the people of South

Carolina. On November 22, 1823, Taylor wrote a petition contesting the election of Thomas P. Davis to the senate of South Carolina.[85]

John Man Taylor, the confirmed bachelor rice baron of Mansfield, Willowbank and Smerden, died of tuberculosis and malaria on December 19, 1823, at the age of thirty-six. He was buried in the cemetery at Prince George Winyah Episcopal Church the following day.[86] In memoriam, Taylor's epitaph reads:

> *Formed by nature and prepared by education to adorn the councils of his country. He was in the meridian of life withdrawn by his God from all earthly scenes of usefulness and honor. The firmness of his character endeared him to the brave. His integrity associated him with the pure in heart. He was frank and his friends loved him. The tears of affection and friendship have consecrated the spot where his ashes repose.*[87]

At the time of John Man Taylor's death, rice land had become increasingly more valuable, and the prospect of becoming a planter by any way other than birthright was very slim. By the mid-1820s, choice rice lands sold for upward of $300 per acre, and average-quality tidal flooded river land and inland swampland (which normally produced 2,400 pounds of clean rice) sold for about $100 per acre, whereas inferior rice land sold for about $50 an acre. The best uplands, which would bring 150 pounds of cotton per acre, sold for about $10 to $15 an acre, and inferior pine lands sold for about $0.25 an acre. In 1825, Robert Mills, South Carolina's first state architect, recalled, "When buying a plantation at this time, it was customary to only sell the estate for the value of the rice land."[88]

When Almira Coffin of Buxton, Maine, visited the area during the antebellum period, she confirmed the manner in which plantations were valued and sold: "In purchasing a rice plantation they only count the rice land—the high land and buildings are thrown in! This high land is not valuable for tillage as it is too sandy; but by manureing the corn fields and gardens they manage to raise enough for their consumption."[89]

With the passing of John Man Taylor, the ownership of Mansfield and his entire estate passed to his sister Anna Maria Taylor. In November 1816, Anna Maria Taylor married Reverend Maurice Harvey Lance, the son of Charleston attorney Lambert Lance and Sarah Lance.[90]

Together, the couple bore at least four children, two of whom lived to maturity. Mary was the first child born to the marriage. Anna gave birth

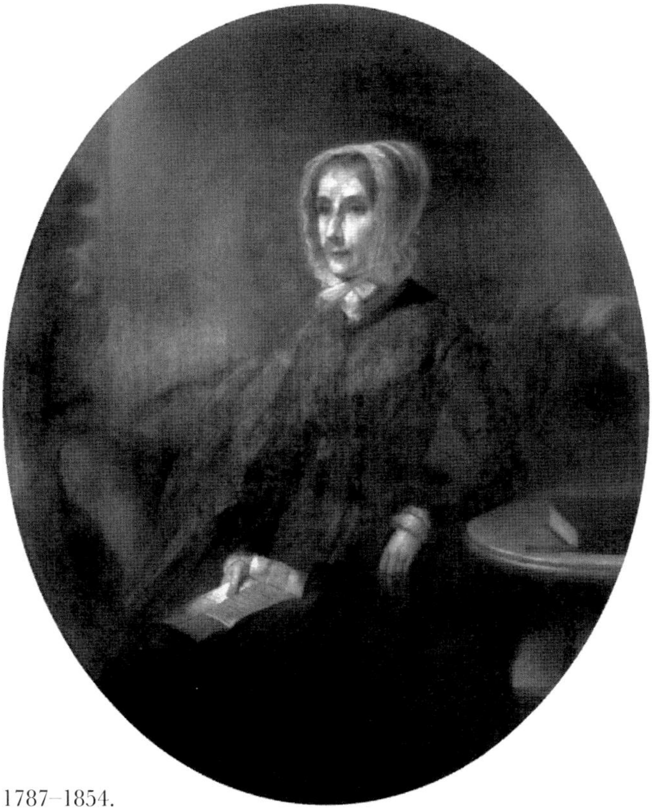

Ann Maria Taylor Lance, 1787–1854.

to her on Easter Sunday, March 22, 1818, and baptized her at Prince George Winyah Church on April 30, Ascension Day, 1818. (Mary's uncle John Man Taylor sponsored her at the baptism.) On May 28, 1819, Mrs. Lance bore Lambert, who was baptized along with three other children at Prince George Winyah Church. The boy lived for only a little over seven months, and the family buried him on January 16, 1820. On September 28, 1820, Anna bore Sarah Lance, and the family baptized her on North Island on August 12 of that same year. Unfortunately, like her brother Lambert, Sarah died in infancy. The family buried little Sarah on August 28, 1822, at one year and eleven days old. After losing two consecutive children, the Lance family felt blessed to have another daughter when Ester Jane was born on June 18, 1822. Shortly after the birth of Ester Jane and the death of John Man Taylor, the family moved to Mansfield.[91]

According to his will, John Man Taylor owned 12,709 acres, town property and 143 slaves in Prince George Parish.[92] As per his wishes,

Mrs. Lance received her brother's rice empire "for and during time of her natural life without being subject to the contract or control of any husband." Taylor entrusted his property and slaves to his sister and made sure that they followed his family blood line by adding the following stipulation to his will: "After her death (Anna Maria Taylor Lance) all money or properties will be held in trust for Mary Taylor and Ester Jane to have, take and receive rents issued and profits of my estate, real and personal."[93]

During the late Taylor occupancy and throughout the Lance occupancy of Mansfield, the planting communities developed into a tightly knit aristocratic faction, while the people in the town became caught up in many of the reforms that swept the nation as a result of the religious revival known as the Second Great Awakening. On December 13, 1817, the women of Georgetown formed the Ladies Benevolent Society, a private organization designed to perform good deeds. In 1827, the townspeople and the clergy (including Reverend Lance) of Georgetown formed the Winyah Anti-Dueling Association to suppress the act of dueling, a favorite way of solving disagreements between planters. In May 1832, the Methodists organized the Winyah Temperance Society (an early crusade against alcohol use). However, the Episcopal Church, the church of the planter class, did not support the society.[94]

The functions of the reform organizations were not an attempt to overthrow the planter class or the social order that they dictated but rather to stabilize life in the town and throughout the countryside. The townspeople moved toward reform measures like the rest of the country; however, they never attempted to form an abolitionist society. Nonetheless, the planter class must have considered the attacks on alcohol and dueling as insults since their code of chivalry promoted dueling and their social lives permitted alcohol use.

While the townspeople became increasingly more in tune with the tide of reforms sweeping the nation, the planters retreated further into their own world. They sipped their homemade and imported wines, worked on their skills in the arts of conversation and politics and relaxed in a lifestyle of ever-increasing leisure. They rocked away on their verandas and in their personal libraries, reading such works as governor of South Carolina and Georgetown resident John Lyde Wilson's *The Code of Honor; or, Rules for the Government of Principals and Seconds in Dueling* (which was recognized as law in the settlement of personal differences between gentleman), Arch Ruport's *The Art of Cockfighting* and the pro-slavery arguments of

Thomas Dew and Iveson Brookes, who countered the rising abolitionist movement in the northern and western states by arguing that slavery was good for the Africans because it brought them Christianity and helped them to rise from tribal to civilized men.

In 1832, most of Georgetown's rice planters aligned themselves with John C. Calhoun in attempting to nullify federal tax laws (declaring unfavorable tax laws unconstitutional within the state) in what became the nullification crisis. After the crisis passed, the planters began to form their own social organizations. The planters who lived south of the town strengthened their already existing ties with the Charleston elite, while the planters north of town formed their own social societies such as the Hot and Hot Fish Club on the Waccamaw Neck, the Planters' Club on the Pee Dee and the Winyah and All Saints Agricultural Society, which also met at the Planters' Club site. Reverend Lance belonged to both the Planters' Club and the Winyah and All Saints Agricultural Society.

The Hot and Hot Fish Club and the Planters' Club were nothing more than male-only convivial clubs for the planters that gave them an opportunity to drink fine wine, enjoy well-prepared dinners and talk politics, but the Winyah and All Saints Agricultural Society was more important, as it encouraged planters to discuss improvements to rice agriculture and related farming matters.

In 1836, the Georgetown "Mission to the Slaves" expanded to the Waccamaw Neck, where a philosophy of improving the lives of the slaves by bringing them Christianity began. Reverend Maurice Harvey Lance was probably responsible for a similar mission in the Prince George Winyah diocese.[95] One Georgetown merchant recalled that Reverend Lance was a "leader in every social and religious enterprise of the district and parish of Georgetown." The merchant went on to explain Lance's life at Mansfield thusly: "His plantation was, for many years, the intellectual center of the life of Georgetown."[96]

Before inheriting John Man Taylor's estate, Maurice Harvey Lance was not a wealthy planter. His estate of 1820 reveals that he owned thirty-nine slaves, of whom only twenty-one were involved in agriculture.[97] He might have been liquidating his estate, as at least one record proves that Lance was selling slaves at this time. On November 27, 1823, Lance sold a mulatto slave named Moses.[98] His tax return for 1824 shows him owning only sixteen slaves in Prince George Winyah Parish.[99] By the 1830s, however, things improved for Lance; with the addition of the Taylor property to his holdings, Lance joined the ranks of the big planters.

The 1830s, which coincided with the Lance tenure at Mansfield, saw an agricultural revolution sweep through the United States.

New and soon-to-be indispensable inventions such as Cyrus McCormick's mechanical reaper in 1831; John Avery and Hiram Abial Pitt's wheat thresher machine, which separated grain from chaff, in 1833; Hiram Moore's combine harvester for grain in 1834; and John Deere's steel plow in 1837 enabled the family farms to become more profitable by both minimizing labor and allowing each farmer to earn greater profit.

Just as small farms became more profitable due to new and improved tools, new inventions allowed the Lowcountry rice planters to become proprietors of industrial plantations. Through trial and error with different methods of power, and finally with the addition of steam power, the process of preparing rice for market became less laborious and more profitable than ever before. With fewer man-hours needed to prepare grain for market, slaves were able to both enjoy more leisure time and also diversify their skills to allow them to perform other tasks on the estate.

The Ravenel family of Charleston constructed a tidal-powered rice mill at the cost of $3,500. This mill did not use the dangerous 360-pound pestles that often damaged or broke the grains but rather used the "rapidity of friction" to clean the grain. Apparently, this new method could be employed to older, preexisting, animal-powered mills and allow planters to send a "brighter" product with fewer damaged samples to market. The editor of the *Southern Agriculturist and Register of Rural Affairs* wrote that the new mill design allowed planters to get rid of the

> *enormous pestle-shaft and lifters, and the heavy pestles weighing 360 lbs; with the expensive foundation of solid brick work required for the mortars to lie upon; a part of the machinery requiring very considerable experience to manage; always attended with danger; propelled (even by water) with difficulty, and hence inapplicable to mills moved by animal power, without sacrifices almost ruinous. For these are substituted pestles moved by cranks, which rise and fall only a few inches, and of size not exceeding 3 x 4 scantling, or about four feet in length.*[100]

Another improvement to rice agriculture during this era came in the form of the rice-threshing mill. Prior to its introduction, threshing was done with the flail stick, an African method of removing the grain from the straw. According to the editor of the same magazine, "threshing, which all must allow, when the crop is a heavy one, is a tedious and laborious operation, is almost universally performed by the simple flail. No machine therefore, can, at the present time, be more desirable than the threshing mill on our plantations."[101]

Frederick Law Olmsted observed and later explained the process by which rice was prepared for market using both the traditional flail stick and winnowing house methods and the steam-powered rice-threshing mill during his visit to the Lowcountry:

> *Threshing commences immediately after harvest, and on many plantations proceeds very tediously, in the old way of threshing wheat, with flails, by hand, occupying the best of the plantation force for most of the winter. It is done on an earthen floor, in the open air, and the rice is cleaned by carrying it on the heads of the Negroes, by a ladder, up on to a platform, twenty feet from the ground, and pouring it slowly down, so that the wind will drive off the chaff, and leave the grain in a heap, under the platform. But on most large plantations, threshing-machines…driven either by horse-power or by steam-power, have been lately adopted, of course, with great economy. Where horse-power is used for threshing, the wind is still often relied upon for removing the chaff, as of old; but where steam-engines are employed, there are often connected with the threshing-mill, very complete separators and fanners, together with elevators and other labor-saving machinery, some of it the best for such purposes that I have ever seen.*[102]

There are several patented rice-threshing machines. Benjamin S. Hort of Georgetown patented the first in February 1812. It was wind powered and thus was subjected to weather conditions. Elias B. Bort of Charleston patented the second machine. Bort's device was animal powered and capable of threshing about two to three hundred bushels of rice per day. Calvin Emmons patented his animal-powered threshing machine in April 1828. A year later, on July 27, 1829, Emmons patented a steam-powered threshing mill. Besides the addition of steam power, another great improvement to his second machine was the use of steel teeth, two inches long and of three-eighths rod, flattened and hardened at the outer ends, and steel plates in concave beds. According to Calvin Emmons, "By steam power, the machines have each threshed 700 bushels per day, though ordinarily from 450 to 500 bushels per day each, and by horsepower, much less."[103]

By this time, too, engineers were working on steam-powered pounding mills so that planters did not have to rely on animal power, winds or tides. On the eve of the Civil War, Georgetown had eleven rice-pounding mills.[104] In 1850, Robert F.W. Allston described the workings of the steam-powered rice-pounding mill at Waverly Plantation:

> *By steam power, the rough-rice is taken out of the vessel which freights it, up to the attic of the building—thence through the sand screen to a pair of (five foot wide) heavy stones, which grind off the husk—thence into large wooden mortars, in which it is pounded by large iron-shod pestles (weighing 250 to 350 pounds), for the space of some two hours, more or less. The rice, now pounded, is more elevated into the attic, whence it descends through a rolling-screen, to separate whole grains from the broken, and flour from both; and also through wind-fans, to a vertical brushing screen, revolving rapidly, which polishes the flinty grain, and delivers it fully prepared into the barrel or tierce, which is to convey it to market.*[105]

Reverend Lance had a very successful career in the ministry even before he graduated from South Carolina College. On February 5, 1814, the Society for the Advancement of Christianity in South Carolina promoted Lance, its first candidate for Holy Orders, to be the lay reader at Trinity Church in Columbia. From there, he received his orders to serve at Prince George Winyah.[106] On February 21, 1815, Bishop Frederick Dehon of South Carolina ordained him deacon. Three and a half years later, on November 8, 1818, Bishop Bowen ordained Lance a priest.[107]

While still officially the rector of Prince George for a few more months, Reverend Lance helped to open Christ Church in Greenville. In Greenville, Lance assisted Episcopal reverend Edward Thomas from his first service in the new sanctuary from June 18, 1826, until January 25, 1827. Mrs. Lance was listed among the first to receive Holy Communion in the new sanctuary on July 9, 1826.[108]

During one of his missions to Charleston, Reverend Lance published a small pamphlet entitled *An Address Delivered in St. Paul's Church, Radcliffeborough, Before the Charleston Protestant Episcopal Sunday School Society on Their Twelfth Anniversary, May 24, Being the Tuesday in Whitsun-Week, 1831.* Reverend Lance returned to Georgetown and, on April 19, 1837, helped Bishop Bowen consecrate Prince Frederick's Chapel in the Pee Dee.

Lance held services at Prince Frederick's until 1840 but returned a few years later to hold them once again.[109]

Reverend Lance left Georgetown each summer to be a true itinerant minister after his tenure at Prince Frederick's. Throughout his travels, Lance was personally responsible for opening several Episcopal churches in South Carolina. In 1843, after spending four months getting to know the people of Spartanburg, Reverend Lance and others started a church at Limestone Springs near present-day Gaffney, South Carolina, a popular resort area. At the same time, Lance and others also initiated a ministry in Spartanburg that was established as Advent Episcopal Church. Lance worked tirelessly spreading the Episcopal faith throughout the upstate, an area dominated by Presbyterian, Baptist and Methodist congregations. On August 9, 1843, he helped Bishop Gadsden consecrate St. Thaddeus in Aiken, South Carolina.

In 1844, Reverend Lance's ministry called him out of South Carolina entirely. That summer, Lance traveled with his longtime friend Richard Lathers and others throughout New England meeting several preachers and visiting congregations before vacationing in Newport, Rhode Island. After a short rest in Newport, Lance and company traveled to New York City and then north to Albany. Heading north, his train derailed and flipped over in Westchester. The reverend escaped the accident unscathed, but his travel companion suffered a broken arm.[110]

In 1847, the reverend ministered in Laurens, South Carolina, where he assisted Bishop Gadsden's three-day visitation services marking the consecration of the Church of the Epiphany.[111] After years of service in advancing Christianity throughout South Carolina, Reverend Lance settled back in Georgetown, where he was once again active in Prince George Winyah. Lance served as one of the wardens beginning in 1848 and throughout the war years.

As a planter, Reverend Lance owned Mauricena Plantation, Jessamine and Woodstock on the Sampit River and Springfield and Smerden Plantations on the Pee Dee, as well as Mansfield, Wedgefield and Greenwich on the Black River at different times. At Smerden alone, Lance's overseer and laborers produced 326.0 barrels of rice in 1835, 275.5 barrels in 1836, 275.5 in 1837 and 244.5 in 1838. S. King served as overseer in 1835–36, and Benjamin M. Grier served in that capacity in 1837 and 1838.[112]

Like their mother, both Lance girls married while they were very young, and married men with bright futures. In 1836, Mary, at the age of eighteen, married Dr. Francis S. Parker of Goose Creek, a graduate of the College of Charleston and Charleston Medical College. Four years later, in 1840, also at the age of eighteen, Esther Jane (who still lived with her parents when the census was taken earlier that year) married Harvard-educated John Harleston Read II. Read, born at Rice Hope Plantation in Moncks Corner, South Carolina, was a politician and rice planter.

Read represented Georgetown in the South Carolina House of Representatives from 1844 until 1861 (Thirty-sixth through the Forty-fourth General Assemblies), when he joined the war effort.[113] As a planter, Read was heir to an enormous rice empire on the Sampit and Pee Dee Rivers.[114] Read eventually became the seventeenth-largest slaveholder in the United States, with a total of 511 laborers.[115] He often leased his slaves to Reverend and Mrs. Lance for forty dollars each per month during the planting, growing and harvest seasons.[116]

On February 3, 1840, Reverend Lance improved the size of his slaveholdings by purchasing a large gang from John Wragg. The Wragg family was liquidating their holdings in both land and slaves at this time and committed to sell forty-four slaves to Reverend Lance for the sum of $29,700. Lance put down a total of $14,850 and financed the rest of the money. The Wragg family also sold several slaves to the Mills, Read, Magill, Gibbs and Gadsden families at this time.[117]

In 1841, Reverend Lance and his wife, Anna Maria Taylor Lance, traded Greenwich Plantation and a sum of money to their son-in-law Dr. Francis S. Parker for Wedgefield Plantation. Then, against the wishes of Reverend Lance, Anna Maria had her estates appraised and gave Dr. Parker and her daughter Mary the estate adjoining Wedgefield "and all the slaves and stock" upon the estate, which were appraised for $62,134.[118]

Also against the wishes of Reverend Lance, Anna Maria transferred ownership of Smerden Plantation "and all slaves and stock" upon the estate, as well as some Sandy Island tracts, valued at $57,254, to her daughter Ester Jane upon her marriage to John Harleston Read II.[119] The Reads eventually renamed Smerden Plantation Belle Rive Plantation (meaning "beautiful shore" in French) after their eldest son, John Harleston Read III, began to attend Belle Rive School on the banks of Lake Geneva in Vevey, Switzerland. Besides Belle Rive on the Pee Dee River and the Sandy Island tracts, the Reads' rice empire eventually included the enormous Maryville Plantation (which Read inherited when his father died in 1859 and which his father had expanded by purchasing the neighboring Oakley, Lucerne and Upton Plantations) at the mouth of the Sampit River, Harmony Plantation (which he inherited after his brother's death) on the same river, Cheraw Swamp Plantation on the Pee Dee, a summer home in Plantersville and a home in Charleston.[120]

Anna Maria stipulated that the properties and slaves were to be transferred to her daughter's children, not to their husbands, in the event that the women died before their husbands.[121] The Lance girls, just like their mother and most southern women, traditionally turned the management of their plantations over to their husbands.

What seemed to be a gesture of goodwill to help Anna Maria Taylor Lance's daughters achieve financial stability started a series of lawsuits initiated by Reverend Lance against his wife and the Parkers and Reads; later, after the death of Anna Maria Taylor Lance, there were more suits and countersuits between the Parkers and Reads.[122]

The lawsuits and financial stress might also be the reason why Reverend Lance left his congregation at Prince George Winyah to become an itinerant preacher. By 1850, the population census revealed that Reverend Lance and his wife resided at different addresses. While Anna Maria called Wedgefield Plantation home, Reverend Lance took up residence on the Pee Dee River.

The Wedgefield estate, which Dr. Francis S. Parker traded to Reverend and Mrs. Lance, consisted of 399 acres, a manor house, typical plantation outbuildings and a pounding mill, which Reverend Lance used to pound his own grain and mill the rice of his neighbors. The average milling fee for a rice crop was 9 percent of the total crop milled at this time, but Lance charged only 8 percent (meaning that he received 8 percent of the grain for his milling service). In 1844, in an apparent attempt to earn money, Lance advertised that the mill was "in good order to beat on toll." Lance also provided barrels at seventy-five cents each.[123]

From the advertisement, it is clear that Reverend Lance's rice-pounding mill was built on the old models of both being tide operated and also still using the heavy pestles to pound the grain. Tidal-powered rice mills could expect to be able to pound 120 barrels of rice during days of common tides and about 170 barrels of rice during days of spring tides. At this rate, the mill could probably mill an average of 132 barrels per day during the eleven-month growing season (allotting one month for harvest). With 252 working days (taking 41 days off for Sundays), Lance could expect to earn between 450 and 500 barrels of rice for his milling services, which would allow him to earn about $7,500.00 per year, and another $337.50 to $375.00 for barrels for the grain, in addition to his own planting. Plus, all of the tailing and chaff removed from the grain could be milled into enough rice flour to feed hundreds of sheep, cows and pigs or could be sold for additional income.[124]

During the Lance residence at Mansfield, between 1830 and 1840, the total number of persons in Georgetown District had decreased, yet the percentage of slaves in the community remained about the same. In 1830, the population of Georgetown was 19,943, of whom 17,798 were slaves. However, by 1840, of the 18,274 persons living in the Georgetown District, 15,993, or 88 percent of the total, were slaves.[125] This phenomenon is due to the fact that the larger planters bought out many of the smaller planters, who then proceeded to acquire new lands in the southwest. In the district as a whole, the merchants continued to lose influence to the Charleston merchants and to the local planters who operated north, south and west of the town.

## *Chapter 3*

# Dr. Francis S. Parker, Rice Planter

*Gentility is not wealth; it is not in dress, nor genius, nor any other attribute of man, unless it goes hand in hand with a cheerful disposition, an amiable temper and a philanthropic will. There are real bonafide gentlemen in all classes of society, and he is the greatest gentleman who mingles with all, and remains untarnished.*
—Pee Dee Times, *May 14, 1856*[126]

By the time Dr. Francis Simons Parker relocated to Georgetown, he was already South Carolina royalty. Parker was born on March 4, 1814, to John Parker V of Charleston and Hayes Plantation in Goose Creek and Emily Smith Rutledge Parker, daughter of General John Rutledge and granddaughter of South Carolina governor John Rutledge. His family ties to South Carolina's history did not end there, however. His great-grandfather had signed the Constitution of the United States, and two grand-uncles were signers of the Declaration of Independence. With such a pedigree, Parker seemed preordained for greatness and distinction in his home state. His family had always served their state, and he, too, would be ready to serve South Carolina when called on by her constituents.

By the 1840s, Georgetown was the premier location for rice planting in the United States, and Dr. Parker moved to the region to cash in on its prominence. His rice-planting career in Georgetown began with the purchase of Wedgefield Plantation (perhaps named for the shape of

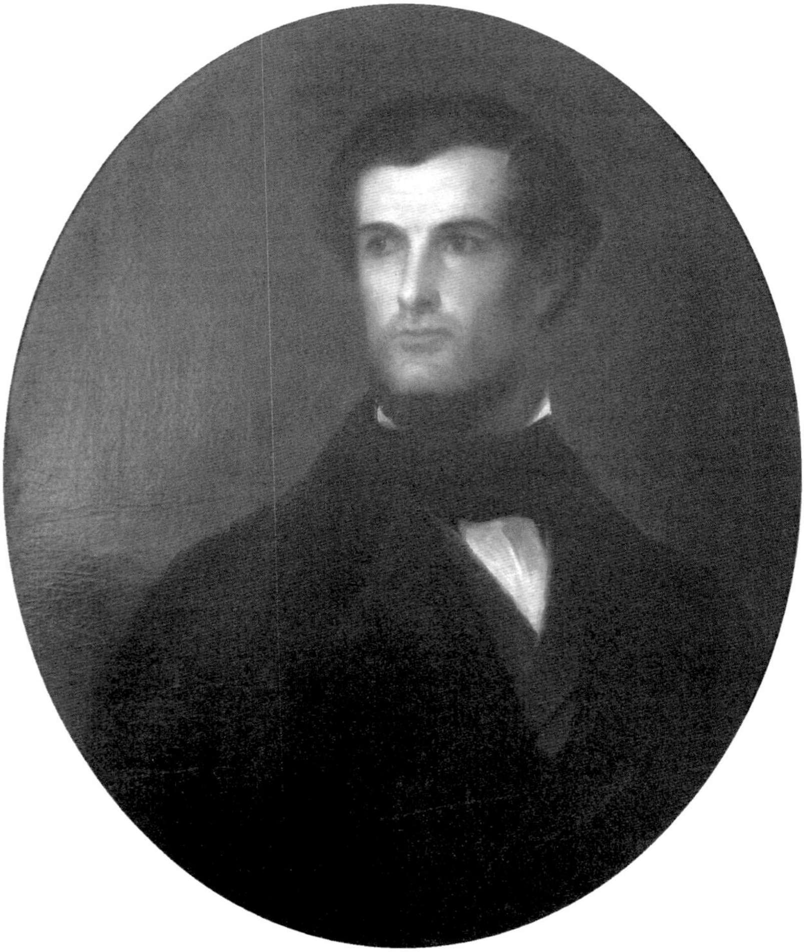

the plantation) on the Black River from Samuel Wragg in 1840. Soon thereafter, Parker purchased Mitchell's Island.[127]

After harvesting only one crop at Wedgefield, Parker traded it to his in-laws for Greenwich Plantation and a sum of money.[128] Later, in 1841, Parker's in-laws gave him and his wife, Mary, Mansfield Plantation and 128 slaves.[129] The Parkers made Mansfield their primary home. At least one of their children (Julia Rose, born on November 29, 1843) married at the estate, and one of their sons (James Rose, born on October 25, 1848) was baptized there.[130]

During Parker's first year at Mansfield, his gang planted 220 acres of "No. 1" quality rice land and 50 acres of "No. 3" quality rice land. He

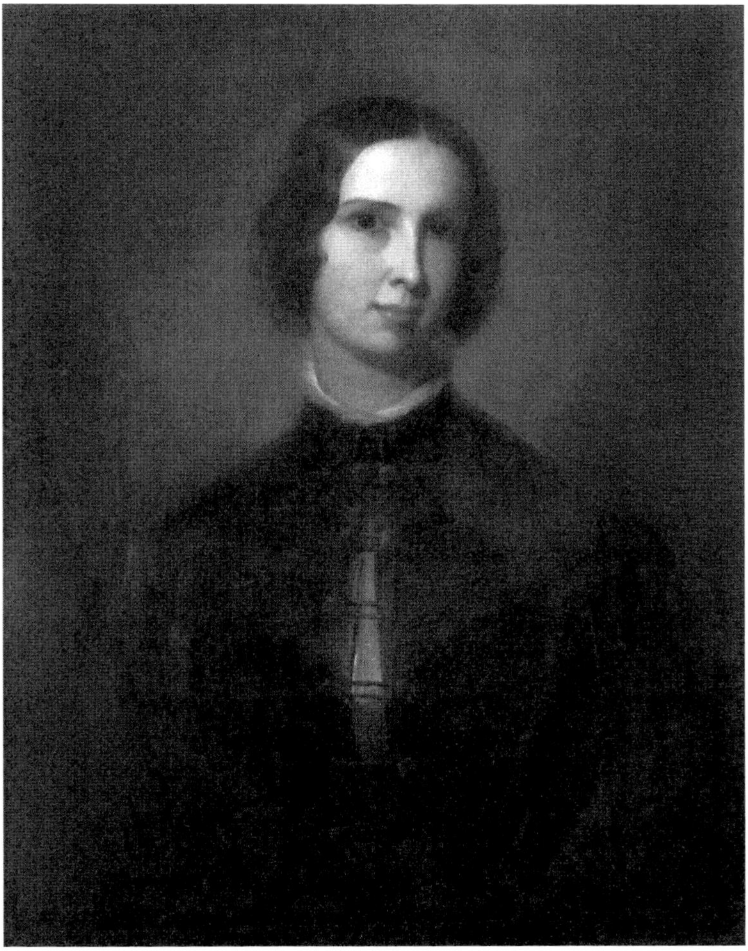

*Above*: Mary Lance Parker, 1818–1888.

*Opposite*: Dr. Francis S. Parker, 1814–1867.

cultivated 50 acres of highland in vegetables and also owned 412 acres of pine barren at the Mansfield tract.[131] Although they later added other estates to their holdings, Mansfield remained the seat of Dr. and Mary Parker's personal rice empire.[132]

Shortly after taking possession of Mansfield, Parker took inventory of his tools and listed them in his journal. The list included: eighteen large rice hoes, six spades, eighteen small rice hoes, eight axes, one pair mill stones and spindle, three mule collars, six scoop spades, three pairs of trace chain, two pair of harnesses, twenty-four broad hoes, twelve trenching hoes and one ox chain.[133]

When Dr. Parker entered the rice-planting profession, the construction of the Georgetown rice culture was nearly complete. Over 160 plantations had been built on Georgetown's intricate river system, the slaves were well disciplined and a distinct class of aristocracy had emerged.[134] During that time, the South Carolina Lowcountry entered into a period of unprecedented wealth; however, the wealth was nearly entirely concentrated in the countryside, and very little was spent on internal improvements such as building up the town, roads and bridges. One visitor to Georgetown in early the 1840s remarked:

> *Georgetown District is the wealthiest portion of the State; but a more miserable collection of decayed wood domiciles and filthy beer shops than are clustered together to make up the town, would be difficult to find. Indeed, unlike the free states, the wealth of the South lies almost entirely in the country; the towns, unless Charleston forms an exception, being made up of artisans and traders.*[135]

As the planters grew wealthier, produced larger quantities of rice and became more influential, they became more closely linked to Charleston, and some even ignored Georgetown altogether. Charleston possessed large-scale rice-milling operations and many factors and bankers who processed and marketed the annual rice crops both nationally and internationally.

The local planters had, since colonial days, been interested in improving their agricultural production. According to tradition, rice and indigo planters formed a loosely knit convivial club that met monthly at the Old Oak Tavern on Bay Street. Soon thereafter, the group began to call itself the Winyah Indigo Society. In 1758, the organization received a royal charter from King George II and quickly became a philanthropic society whose main concern shifted from agricultural improvement to housing and educating orphans and paying the salary of the orphans' teacher.[136]

The area's next attempt to develop an agricultural society was the formation of the Planters' Club on the Pee Dee on November 8, 1839. The organization, unlike the Hot and Hot Fish Club of Waccamaw, to which many of the members belonged, was open exclusively to rice planters. The club quickly became a convivial club, and in June

1840, the South Carolina Agricultural Society refused the Planters' Club's charter.[137]

Then, in November 1842, Georgetown rice planters formed the Winyah and All Saints Agricultural Society. Besides the Allston, Tuckers, Heriots, Sparkmans and other big planters of the era, Dr. Francis S. Parker; his father-in-law, the Reverend Maurice Harvey Lance; his brother-in-law, John Harleston Read II, who served as the society's secretary and treasurer; and eventually Parker's two eldest sons, John and Frank Parker, were members of the club. The organization's stated purpose was to encourage improvements in agriculture; improve the domestic economy; conduct experiments; study animal diseases; and improve the breeding of horses, mules, sheep and cattle in the district. The organization met three times each year at a clubhouse it had built at Guendalos Plantation on the Pee Dee River.[138]

At its November gathering, John H. Tucker, president of the society, gave an address that was largely devoted to politics and southern agriculture. The organization also used its November meeting, as well as its January meeting, to plan its agricultural fair. The society's April meeting, which was regarded as its anniversary meeting, was the organization's finest assembly. The engagement was the occasion for the society's annual fair and cattle show, which offered premiums for the finest examples of crops and livestock. The first premiums were monetary prizes of five dollars; however, by the mid-1840s, the society had begun to present winners with silver medals that had a picture of a sheaf of rice and a plow on one side and the victor's name and category for which the premium was rewarded on the other.[139]

By the 1840s, agricultural societies and their fairs were quite popular throughout the American South. Agricultural fairs had their roots in medieval European commercial fairs, where livestock and agricultural products were commonly sold. By the time the Winyah and All Saints Agricultural Society was formed, there were fourteen agricultural societies in South Carolina. Most of the agricultural societies in the state sponsored fairs that were hailed as great social and community unifying events.[140]

At a typical agricultural fair and cattle show, visitors could expect to see many different crops and various livestock on display. The public was entertained by orations on agricultural and livestock improvements by members of the society, and new inventions were often placed on display, such as machines for draining and flooding rice fields. The society also held an auction at the fair in which exotic livestock, such as rare breeds of cattle and sheep and imported plants, were auctioned off to the highest bidder.[111]

Under the tutelage of the Winyah and All Saints Agricultural Society, the Georgetown rice planters reached their climax of agricultural prosperity and entered into a period of wealth unparalleled in South Carolina history. During the early 1840s and 1850s, the society made its most important advancements in agriculture. The first was the discovery of large-grain rice by Colonel Joshua John Ward; another was Dr. Edward Thomas Heriot's system of fertilizing with rice straw, chaff and flour; and the final was Dr. Francis S. Parker's system of fertilizing with bat guano. By the 1850s, all of the members of the Winyah and All Saints Agricultural Society, as well as rice planters throughout the Lowcountry, were supplied with these scientific advancements. At the January 1860 meeting, which happened to be one of the last meetings of the society, Dr. Parker gave a lecture and report on the use of grain for fertilizer, which the society published and distributed to its members.[112]

During the antebellum period, the *Southern Agriculturist, Horticulturist, and Register of Rural Affairs* was a staple publication. In 1841, J.D. Legare, editor of the magazine, published an article entitled "Synopsis of the Culture of Rice on Black River." Legare suggested two methods of planting rice. One was called open-trench planting. Open-trench planting required that the ground be ploughed, harrowed and broken or dug with a hoe. The ground was then trenched with trenching ploughs to make furrows. Next, the rice was "clayed," meaning it was put inside a clump of clay, watered and then planted. The other form was called covered planting. Covered planting required the same preparation of the fields, but it differed in that rather than being clayed, the seeds were simply broadcast over the field and covered with water. Either way, planting required two and a half bushels of seed per acre, and planters could expect to harvest as much as eighty pounds of rice per acre and forty-seven pounds per bushel.[113]

Like agricultural improvements, livestock improvements were a constant concern of the Winyah and All Saints Agricultural Society. The society imported new types of cattle for itself and for the farmers of the pine forests. For example, the society introduced Brahmin and Tucson cattle to perform field work and Devon cattle for dairy purposes and butchering. The organization also improved sheep livestock in the district; breeds such as the Merino, Southdown, Bakewell and Smyrna, also known was the broad tail, were introduced to the region prior to the American Civil War.[144]

During the 1850s, the Winyah and All Saints Agricultural Society began to sponsor its members in world competitions. In 1851, at the Crystal Palace exhibit of London's World Fair, Dr. Edward Thomas Heriot was awarded a prize medal for his sample of "Carolina Rice."[145] The next year, Joshua John Ward received the highest award for his rice sample at the Paris Exposition, and in 1855, Robert Francis Withers Allston won a silver medal at the same competition for a sample of rice from his Chicora Wood Plantation. Allston's sample was left in a jar for preservation and, under the auspices of the Department of War, was entered into competition the following year at the Algerian exhibit, where it won a gold medal.[146]

Though dominated by rice planters, the organization included doctors, merchants and farmers. Membership in the organization was open to all white males, and the achievements of the society were helpful to South Carolinians throughout the Lowcountry. The Winyah and All Saints Agricultural Society was a very useful organization that provided a powerful communal link between the wealthy rice planters, the Georgetown merchants and the yeoman farmers of the pine forest. The society enabled planters to prosper through shared professional achievements until it closed its doors in the spring of 1861 due to the American Civil War.[147]

Guided by a conscientious and informed agricultural society, and coupled with the inventions that made the rice estates industrial plantations, the 1850s proved to be the most profitable decade for the rice planters as the value of both rice and rice land reached their all-time antebellum highs. During the decade, rice sold from between 2.9 and 4.3 cents per pound. Several planters sold over 1 million pounds each year, netting a minimum of $32,000.[148] Meanwhile, the average salaries of the inhabitants of town were only $1,200 per year.[149] During the 1850s, quality rice land in the district sold for $200 to $300 per acre, as it had in the late 1820s, and undeveloped swampland sold for $100 per acre.[150]

The census report taken in 1850 shows the population of the Georgetown District to be 20,647, 88 percent of whom were slaves. In 1850, only 2,193 of the 20,647 were free persons, including 201 "free persons of color." Of those listed in the census, 604 white persons, 924 slaves and 100 "free persons of color" resided in the town, while the rest were scattered throughout the countryside.[151]

In 1850, South Carolina had 4,072,651 acres under cultivation and 29,969 total farms. Of that total, Georgetown had 49,609 acres of land under cultivation with a cash value of $5,559,100, the highest land value of any district in the state. Since Georgetown planters were so devoted to rice production, Georgetown earned the lowest values of both animals slaughtered and orchards. Georgetown also had a small white population and the third-lowest number of farms (550) in the state. Only Richland (city of Columbia) and Kershaw had fewer farms.[152]

Even though there was great wealth in Georgetown, the district was still one of the most rural of the entire state. In 1850, the state of South Carolina boasted 52,642 homesteads. Charleston contained by far the most homesteads in the state, with over 5,000, while Georgetown had the least amount of dwellings by either white people or free persons of color. Georgetown's 575 homesteads made it the most rural district, followed by Williamsburg with 717, Kershaw with 928, Marlboro with 929 and Horry with 980. All other districts had over 1,000 homesteads, but none other than Charleston, Spartanburg and Edgefield contained more than 2,500.[153]

In 1850, thirty-five-year-old Parker was well on his way to becoming one of the top rice planters in South Carolina and the nation. According to the 1850 agricultural survey, he harvested 375,000 pounds of rice, making him the fourth-largest grower on the Black River. His property included 370 acres of improved land and 1,500 acres of unimproved land with $400 worth of farming implements and machinery. Parker valued his livestock at $890, which included four horses, fives mules, ten milch cows, twelve working oxen, eight other cattle and twenty sheep. Besides rice, Parker's fields produced eight hundred bushels of Indian corn, five hundred bushels of oats, three hundred bushels of peas and beans and three hundred bushels of sweet potatoes. In 1850, Parker's plantationlands were estimated to be worth $25,660.[154] According to a separate appraisal prepared for the Charleston law firm of Simons and Simons in 1850, Parker's gang at Mansfield alone was valued at $42,125. Individuals ranged in value from lead carpenter Daniel, who was worth $600, down to slaves no longer in service, such as Miley, Sukey and Old William, who were valued at $25 each.[155]

Although Parker was an enormously successful planter, his brother-in-law, John Harleston Read II, who was managing his father's estates as well, had much greater success. In 1850, Read's plantations were valued at $90,000. Upon these lands, Read produced 960,000 pounds of rice with the use of 323 slaves. At Belle Rive alone, Read worked 365 acres of improved land and held 595 acres of unimproved land. The improved acres yielded 330,000 pounds of rice, 850 bushels of corn, 160 bushels of oats and 1,100 bushels of sweet potatoes. His livestock on hand at Belle Rive included eight horses, six mules, six milch cows, fourteen working oxen and forty-four sheep.[156] From 1847 to 1853, Read also owned a one-third share in the *Charles Kershaw*, a steamship used to transport rice to factors in Charleston.[157]

At that time, Parker and Read's father-in-law, the Reverend Harvey Maurice Lance, made Wedgefield Plantation and his other landholdings more profitable enterprises as well. Like Parker, Lance cultivated 370 acres of improved land but had 1,000 unimproved acres, about 500 less than Parker. The value of Lance's machinery was rated at $500, perhaps because he had a rice-pounding mill whereas Parker had only a steam-powered threshing mill. Lance's livestock, valued at $810, included three horses, five mules, fourteen milch cows, ten working oxen, ten other cattle and twelve sheep, from which he produced thirty-six pounds of wool. Lance, too, produced a few other agricultural staples in addition to rice.

Lance produced 420,000 pounds of rice on Wedgefield, Willowbank and Mauricena with the work of 262 laborers. The same laborers produced 300 bushels of oats and 1,200 bushels of Indian corn, but unlike Parker, Lance failed to grow sweet potatoes, peas and beans. Lance's entire agricultural settlement was valued at $24,000 in 1850.[158]

On October 27, 1851, Anna Maria Lance died. According to John Man Taylor's will, the properties that she inherited from his estate were to be divided among her daughters. Complicating matters, in February 1844, Mrs. Lance made her daughter Mrs. Ester Jane Lance Read the trustee of John Man Taylor's will.[159] Each of Lance's two sons-in-law was poised to receive a handsome inheritance from the estate; however, they disagreed over how much each should receive.

Reverend Lance immediately balked at the proceeding of the will. Richard Lathers, a Georgetown merchant and longtime friend of Reverend Lance, recalled the family legal battles: "The wealthy relatives of Mr. Lance, having a baseless prejudice against Mr. Lance, constructed, with the aid of one of the most astute legal lights of Charleston, a trust will in favor of Mrs. Lance and their children, by which Mr. Lance was not only cut off from any direct participation in the estate, but by which he was prevented from exercising any supervision over it."[160]

After eliminating Reverend Lance from the inheritance, Mr. Read and Dr. Parker opened a new round of arguments and lawsuits over the land, slaves and money willed to their wives. According to Mrs. Lance's will, the Parkers received Willowbank Plantation.[161] On January 10, 1852, the Reads opened a round of legal suits between the two sisters and their family inheritances. The Reads sued the Parkers because Mansfield had appraised higher than Smerden (by this time renamed Belle Rive) at the time that they received the plantations from their mother a decade earlier. At the conclusion of this case, the Parkers were ordered to pay the Reads $4,890. A month later, on February 2, 1852, the Reads sued the Parkers for the slaves who lived at Willowbank. The court decided in this case that the slaves should be divided equally between the two sisters. In June of the same year, Mary Taylor Parker countersued the Reads for the slaves at Willowbank and won the case. In November and December 1852, the Reads won two cases, which gave them the rights to sell the Georgetown city plots of land that Mrs. Lance owned as part of the estate of John Man Taylor and to keep all of the proceeds from the sale. Finally, on January 30, 1854, the court reversed its decision of June 1852, which gave all of the Willowbank slaves to the Parkers, and once again split the gang between the Reads and Parkers.[162]

In 1854, the reverend married Sarah Laura Smith at St. Paul's Church in Charleston. Sarah was the youngest daughter of Samuel Smith of Georgetown. Smith was a small farmer who never owned more than ten slaves but, upon his death, transferred his holdings to his daughter Sarah Laura.[163] The newlyweds returned to Georgetown and continued to plant rice.

By the end of the decade, Dr. Parker had become one of the largest rice planters in America. In 1860, his production had nearly quadrupled his 1850 production figure, and he out-produced his ex-father-in-law by 100 percent. He harvested 1,444,000 pounds of rice, the second-highest yield on the Black River. According to the Agricultural Survey for 1860, Dr. Parker had 429 acres of improved land in production and 2,250 acres of unimproved lands as well. By that time, the value of his farm implements was $10,000, and the cash value of his farms had almost quadrupled since 1850 to $100,000. Parker's livestock was valued at $3,150; he owned six horses, fifteen mules, ten milch cows, eight working oxen, four other cattle and fifty sheep, from which he produced wool. Parker also produced three thousand bushels of Indian corn, thirty bushels of peas and beans and two thousand bushels of sweet potatoes.[164]

During the 1850s, John Harleston Read II expanded his rice empire as well. First, he purchased Cheraw Swamp Plantation on the Pee Dee from Francis Weston, which included "584 acres of tide swamp, 180 of which is cleared and under cultivation and 101 acres of upland on Sandy Island." In 1856, he inherited the properties of his only sibling, James Wither Read, after James's death. And after his father died in 1859, Read inherited Maryville Plantation on the Sampit River. In 1860, Read's enormous rice empire produced 1,575,000 pounds of rice by the hands of 511 slaves.[165]

Reverend Maurice Harvey Lance, too, expanded his holdings during the decade but did not have the same results. In 1860, Lance produced 560,000 pounds of rice on his 450 acres of improved land with 1,000 acres of unimproved land. The value of his farming implements and machinery was still only $500, but the value of his farms had increased to $30,000. Lance possessed $85,000 worth of personal property, while his forty-five-year-old wife, Sarah, retained the rights to an additional $6,000 worth of personal property.[166] He owned four horses, eight mules, six milch cows, eight working oxen, twenty-five other cattle and fifty sheep, which produced 100 pounds of wool, for a total of $1,500 worth of livestock. His supporting crops produced at Wedgefield and other estates included 1,500 bushels of Indian corn and 250 bushels of oats.[167]

On the Black River, four families—Pringle, Trapier, Middleton and Parker—dominated rice production. John Julius Pringle of Greenfield and White House grew 150,000 pounds of rice in 1850 but had increased that total to 720,000 pounds by 1860. The Trapier family increased its production of 960,000 pounds in 1850 to 2,100,000 pounds in 1860, Henry A. Middleton increased his yield from 900,000 pounds to 1,300,000, and of course, Dr. Parker increased his quantity of rice production from 375,000 pounds in 1850 to 1,440,000 pounds in 1860.[168]

Several things account for Parker's increase in rice production. Parker industrialized his estate, acquired additional plantations and increased his slave population from 122 in 1850 to 221 in 1860. He also employed a traveling overseer to monitor the performances of his resident overseers at each of his plantations. Parker also experimented with new fertilizers and other products to stimulate productiveness.[169]

Because the land on the Black River has a lower level of clay and a higher level of humus (decomposed leaf matter) than land on the Waccamaw and Pee Dee Rivers, it was considered inferior to the Waccamaw and

Pee Dee for rice production. However, Parker was able to improve his yield through scientific experiments that allowed Mansfield and his other estates to reach their agricultural potential.[170] As a member of the Planters' Club on the Pee Dee and the Winyah and All Saints Agricultural Society, Parker had access to scientific rice-planting techniques. He took the advice of his fellow society members and began planting golden, big-grain rice and experimented with different manures.[171] First, he tried fertilizing with rice straw and chaff, but he found that guano (bat dung) produced better results.

On July 1, 1859, John Berkley Grimball wrote the following in his diary: "Dr. Francis Parker told me today that he used Guano with marked benefit upon his inferior quality of rice land—that kind called bay land—and in which there is little or no clay. He applies the guano mix with clay at the rate of one ton of Guano to 15 acres. He applies it in the month of June." Apparently, Grimball found Parker's use of guano to be effective on his own property. On August 18 of that same year, Grimball wrote:

> *Fine weather continues and I suppose they are cutting rice today. Mr. Charles Heyward, whom I saw in the cars yesterday, tells me that he finds the manipulated Guano very good for corn—applied when the earth is ploughed away from the corn. Work in the trench on each side and immediately cover up with the hoe or plough—about 150 lbs. to the acre. The Guano which I have had applied to the Rice, is conspicuously beneficial. I shall* [try] *80 to 100 acres in the same way next year. Dr. Parker tells me he has made his land yield half as much again as they did before, by use of Guano.*[172]

Besides the use of bat guano and various other fertilizers, Parker's journal reveals that he attempted to "jump start" his rice plants early with lant emit. Lant emit not only helped seeds get started early but also protected plants from the harmful rays of light. The practice of transplanting small rice plants to the field had already been in practice for at least fifteen years by the time Parker noted the use of lant emit.[173] Thomas Pinckney explained how the transplanting process compared to the traditional common planting process.

### Planting Techniques Compared[174]

| Common Planting | Days | Transplanting | Days |
|---|---|---|---|
| *Trenching, sowing and covering at ⅓ of an acre per hand* | *3* | *Sowing seed bed and drawing plants* | *1* |
| *Bird minding at ½ acre* | *2* | *Trenching or marking the ground for transplanting at ½ acre* | *2* |
| *Two hoeing at ½ acre* | *4* | *Transplanting at 1⁄10 of an acre* | *10* |
| *One hoeing at ¼ of an acre* | *4* | *One hoeing at ½ an acre* | *2* |
| *Hand picking at ⅓ of an acre* | *3* | *Hand picking at ½ an acre* | *2* |
| *Total* | *16* | | *17* |

Another interesting entry in Parker's ledger is the purchase and use of work dogs. The *bouvier des flandres*, which translates from the French language to "driver of oxen," has traditionally been known as the "driver dog." These dogs, whose original purpose was to pull carts and drive oxen to the market, helped Parker's slaves by pulling wagonloads of grain, thus saving them from performing strenuous manual labor. The dogs are naturally territorial and were also used as herding dogs.

In addition to using different fertilizers and dogs, Parker improved his production by taking ownership of Hiwassie Plantation by 1850. He also took over management of his father's Hayes Plantation in Goose Creek and, in 1859, purchased Hopeland Plantation on the Santee River from Joseph A. Winthrop, executor of the will of Thomas L. Bulow, who reported growing 210,000 pounds of rice on the estate in 1850 with the help of sixty-two laborers. Parker held the estate for a few years before selling it in 1864 to George A. Trenholm, secretary of the treasury of the Confederate States of America.[175]

The 1860 population census reveals that forty-five-year-old Dr. Francis S. Parker had $125,200 worth of real estate and $130,000 of personal property. Living with him was his wife, Mary (forty-two); John Emilius (twenty-two); Anna Mary (twenty-one); Francis Simmons

Jr. (twenty); Julia Rose (nineteen); Arthur Middleton (fifteen); Emily Rutledge (thirteen); James Rose (twelve); Rutledge (nine); and Daniel Hayes (five).[176]

In the same report, Parker's brother-in-law, John Harleston Read II (forty-three), was listed as owning $202,000 worth of real estate and $204,400 worth of personal property. His family consisted of his wife, Ester Jane (thirty-five); sons John Harleston III (seventeen), James (eight), H.B. (six) and Lance (two); and daughters Ellen (thirteen), Ester Jane (eleven) and Anna (four).[177]

The 1860 census also reveals that South Carolina had 482 farms of 1,000 acres or more (which is the largest size category enumerated) and another 1,359 farms of 500 to 999 acres.[178] The population census for Georgetown included 3,013 whites, 183 "free colored" and 18,109 slaves.[179]

For whatever reason, whether for profit or to get rid of troublesome laborers, Parker sold twenty-three members of his gang to James L. Paul on January 26, 1860. On that same day, perhaps at auction, Parker sold a fifty-five-year-old male slave named Charles and a woman named Hagar and her son Jacob to James L. Paul as well.[180] There are few records showing that Dr. Francis S. Parker sold his slaves, but his journal does list names of slaves and the dates on which they were purchased. For example, Parker purchased ten slaves (a man, three women and six children) from "W.B." on September 15, 1857.[181] Parker kept lists of his gang members in his different plantation journals, and the slaves listed on these bills of sales do not match any of the gangs in his record books, so it is hard to determine who these slaves were and on which estates they labored. As late as August 16, 1860, Dr. Parker offered to purchase a young male slave from James B. Betts in Charleston.[182]

Besides looking at these sales documents, one would have to look fifteen years into Parker's past to find proof that he sold slaves out of his family. On March 3, 1845, Parker, acting as a trustee of Theodore Gaillard, sold a slave named Catherine and her two sons Lewis and Hamilton to Solomon B. Levy. There is an interesting note on the bill of sale. Parker, after giving a thorough medical examination to the slaves, declared the children to be of sound health with the exception of a blemish on Lewis's tongue. He diagnosed Hamilton with asthma.[183]

### Georgetown Slaves According to the Census Reports

| Year | 1820 | 1830 | 1840 | 1850 | 1860 |
|---|---|---|---|---|---|
| Male | 7,851 | 8,381 | 7,589 | 8,748 | 9,143 |
| Female | 7,695 | 9,417 | 8,413 | 9,505 | 8,966 |
| Total | 15,546 | 17,798 | 15,993 | 18,253 | 18,109 |

In 1864, the census bureau released the Agricultural Survey addendum to the 1860 population census. Georgetown's contribution to South Carolina's agricultural status was impressive.

### 1860 Agricultural Survey of South Carolina[181]

| Agricultural Product | Dr. Francis S. Parker | Georgetown District | South Carolina |
|---|---|---|---|
| Improved Acres | 429 | 50,858 | 4,573,000 |
| Unimproved Acres | 2,250 | 309,083 | 11,023,850 |
| Cash Value of Farms | $100,000 | $5,818,000 | $130,652,508 |
| Value of Farming Implements/ Machinery | 10,000 | 616,774 | 6,151,657 |
| Horses | 6 | 811 | 81,125 |
| Asses and Mules | 15 | 661 | 56,456 |
| Milch Cows | 10 | 2,376 | 163,938 |
| Working Oxen | 8 | 1,450 | 23,020 |
| Other Cattle | 4 | 6,611 | 320,209 |
| Sheep | 50 | 4,666 | 233,500 |
| Swine | — | 11,416 | 965,779 |

| Agricultural Product | Dr. Francis S. Parker | Georgetown District | South Carolina |
|---|---|---|---|
| Value of Livestock | $3,150 | $310,710 | $23,931,405 |
| Bushels of Wheat | — | 2,220 | 1,285,631 |
| Bushels of Rye | — | 460 | 89,091 |
| Bushels of Indian Corn | 3,000 | 339,373 | 15,065,606 |
| Bushels of Oats | — | 8,755 | 936,974 |
| Pounds of Rice | 1,440,000 | 55,805,385 | 119,100,528 |
| Pounds of Tobacco | — | 50 | 104,412 |
| 400-Pound Bales of Ginned Cotton | — | 106 | 358,412 |
| Pounds of Wool | — | 7,054 | 427,102 |
| Bushels of Peas and Beans | 30 | 19,270 | 1,728,674 |
| Bushels of Irish Potatoes | — | 3,039 | 226,735 |
| Bushels of Sweet Potatoes | 2,000 | 139,970 | 4,115,688 |
| Value of Orchard Products | — | $310 | $213,989 |
| Gallons of Wine | — | 10 | 24,964 |
| Value of Market Garden Products | — | $100 | $187,348 |
| Pounds of Butter | — | 16,030 | 3,177,934 |

| Agricultural Product | Dr. Francis S. Parker | Georgetown District | South Carolina |
|---|---|---|---|
| Pounds of Cheese | — | 10 | 1,543 |
| Tons of Hay | — | 745 | 87,587 |
| Bushels of Grass Seed | — | 8 | 38 |
| Pounds of Hops | — | 45 | 122 |
| Pounds of Flax | — | 19 | 344 |
| Hundreds of 1,000 Pounds of Sugar Cane | — | 1 | 198 |
| Gallons of Sorghum Molasses | — | 55 | 51,041 |
| Pounds of Beeswax | — | 203 | 40,479 |
| Pounds of Honey | — | 3,199 | 526,077 |
| Value of Homemade Manufacturing | — | 25,120 | 815,117 |
| Value of Animals Slaughtered | — | $38,710 | $6,072,822 |

Just before the South declared independence, Georgetown remained a rural district, and even though Georgetown contained great wealth, the wealth was still concentrated in the hands of a relative few and did not extend into internal improvements for the area. One visitor traveling through the district just a few days after the signing of the Ordinance of Secession in 1860 rode right by Mansfield Plantation and described the town and the area. He claimed that the trip from Georgetown to North Carolina was a difficult one in those days. He stated, "A trip of seventy miles is something of an undertaking in that region, and quite a crowd gathered around to witness our departure, not a soul of whom, I will wager, will ever hear the rumble of a stage-coach, or the whistle of a steam-car in those sandy, deserted streets." Concerning the town's amenities, he wrote, "The town boasts no public conveyance, except a one-horse gig that carries the mail on tri-weekly trips to Charleston. That vehicle, originally used by some New England doctor, in the early part of the past century, had but one seat…I was forced to seek conveyance at a livery-stable. The only livery establishment in the place, kept by a 'cullud pusson' who, though a slave, owns a stud of horses that might, among a people more movingly inclined, yield a respectable income." The visitor, while leaving the town of Georgetown and heading toward Conwayborough on the Conwayborough–Georgetown Highway (now 701), described the shabby elegance of the Georgetown countryside:

> *We soon left the village, and struck a broad avenue, lined on either side by fine old trees, and extending in an air-line for several miles. The road is skirted by broad rice-fields, and these are dotted here and there by large antiquated houses, and little collections of Negro huts. It was Christmas week; no hands were busy in the field, and every thing wore the aspect of Sunday. We had ridden a few miles when suddenly the road sunk into a deep, broad stream, called, as the driver told me, the Black River. No appliance for crossing being at hand, or in sight, I was about concluding that some modern Moses accommodated travelers by passing them over its bed dry-shod, when a flat-boat shot out from the jungle on the opposite bank, and pulled toward us. It was built of two-inch plank, and manned by two infirm darkies, with frosted wool, who seemed to need all their strength to sit upright. In that leaky craft, kept afloat incessant baling, we succeeded, at the end of an hour, crossing the river. And this, be it understood, is traveling in one of the richest districts of South Carolina.*[185]

Another visitor to the area at this time described the port of Georgetown as having lost its importance to the state. He wrote, "The little port of Georgetown, to which the tradition alone remains of its former importance, when the first colonist from England made it their seat of government. Now, it lives upon the necessities of the neighboring planters, and is a dull unhealthy place, containing about two thousand inhabitants. It is, however, conveniently situated on the Winyah."[186]

By 1860, the planting class had strengthened its hold on local influence. In that year, the wealthiest 3.8 percent of the white population owned 50.1 percent of the land in the district. A total of 82 slaveholders in the Georgetown District owned 80 or more slaves for a combined total of 14,638, or over 80 percent of the district total; 399 slaveholders accounted for the rest of the slaves in the district.[187]

Like the planting class throughout South Carolina and the other southern states, rice planters felt confident in building a nation of their own. Like their forefathers who fought a war of independence from England to maintain inalienable rights, to achieve personal rights and freedoms and to maintain local autonomy, the planters led the way to independence and nation building by throwing off the perceived yoke of tyranny and seceding from the United States.

## Chapter 4

# Dr. Francis S. Parker, Physician

*The art of medicine consists in amusing the patient while nature cures the disease.*
*—Voltaire*

Francis S. Parker started his career at a very proper place for South Carolina planters: he graduated from the College of Charleston in 1833 and was the first honor graduate of the Medical College of South Carolina in 1838.[188]

He probably never planned to be a doctor to the public. There were a few professional traveling doctors in Georgetown, but it was common among sons of great planters to receive medical training before they took to managing their own estates so they could provide their own necessary medical services. Of course, some young men did not always listen to their parents. One plantation mistress asked her young son to not attend West Point and instead train in medicine because a medical education had many purposeful applications. She said, "Save yourself the expense of doctor's bills on your plantation, and in your family."[189]

In the Georgetown District alone, at least twenty planters earned their medical degrees between 1820 and 1860. Among those who earned their degrees were Edward Thomas (affectionately known as E.T.) Heriot and one of his sons of Sandy Island, John D. Magill and one of his sons, John H. Tucker's two sons, three members of the Flagg family, Paul Weston of Waccamaw, James Ritchie Sparkman of the Pee Dee and B.F. Trapier and John Asby Wragg of the Black River.[190]

By the late antebellum period, perhaps due to the rising influence of the organized abolitionist movement, the title of "planter" had become tainted and had lost much of the great respect it commanded during the colonial and early antebellum days. In response to this loss of influence, planters who had earned their medical degrees reported themselves as being physicians in the 1850 and 1860 census reports instead of listing their occupations as planter.[191] In addition to the official census reports, planters who were medically trained (such as E.T. Heriot and Robert Nesbit) were memorialized as physicians instead of planters in their planter club records and in local newspapers. In the case of E.T. Heriot, his contemporaries referred to him as a "doctor" because it was "a pursuit more congenial to his views."[192]

Throughout his antebellum ownership of Mansfield, Parker kept notes of the needs and products ordered and consumed by the residents of Mansfield. One of the first notes in his journal is a detailed description of the medicinal requirements and annual supplies needed. These lists, when analyzed, give insightful evidence into the health of the people who lived at Mansfield.

Among the annual supplies ordered by Dr. Parker were two barrels of salt, two barrels of molasses, three hundred bushels of flour and two barrels of pork (bacon) at three different times of the year (Christmas, seed time and harvest). I think that it is important to note that although the planters of the South Carolina rice coast were absent from the estates during much of the year, they were in residence when the pork allotments were given out. They would return at harvest and stay through Christmas before going to Charleston for the "social season" (usually mid-January to Easter) and returning at planting time.

Obviously, allotments were not the only things that either the laborers at Mansfield or the Parker family consumed. The South Carolina Lowcountry was already well known as a land of johnnycakes and waffles, hoecakes and hominy long before the Parkers took residency at Mansfield. Dr. Parker's journal clearly indicates that oats, corn and peas were planted at the estate each year and were probably consumed on a regular basis by the entire labor force in addition to the rice crop. In addition to sustenance crops and allotments, slaves had small gardens behind their cabins where they cultivated potatoes, pumpkins, watermelons, figs, tomatoes, grapes, okra, eggplant and other fruits, vegetables and herbs during their spare time for their own families' consumption.[193]

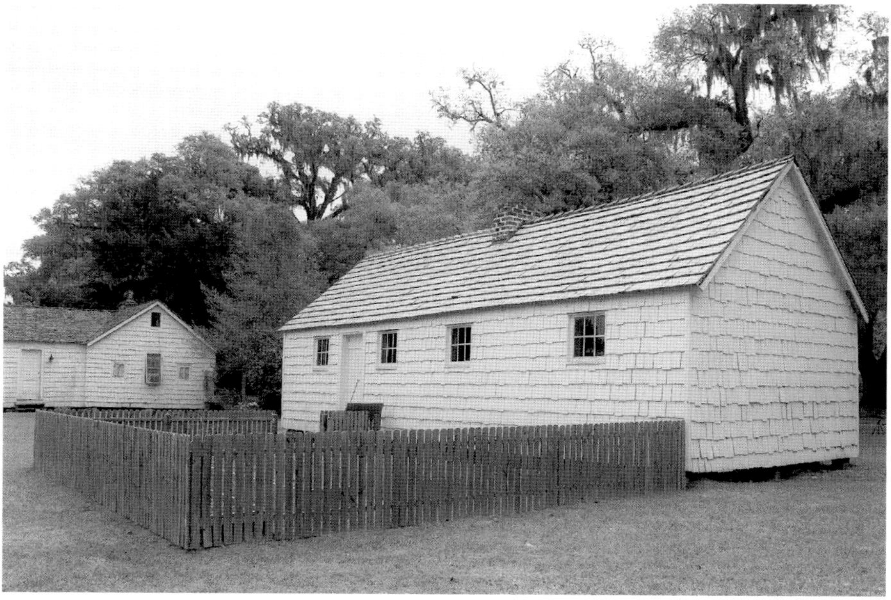

One of the restored cabins on Mansfield's street has an herb and vegetable garden that produces crops reminiscent to what the slaves and, later, freedmen and women cultivated.

Slaves also supplemented their diets on the plantations by hunting and fishing in rivers and sometimes the salt marsh or ocean for oysters, crabs, crawfish and turtles. Professor Jim Michie reported finding lead-impacted shot and gun parts in his study of Richmond Hill Plantation on the Waccamaw River. In the faunal cache at Richmond Hill, Michie recorded finding more fish and wild game such as deer, raccoon, boar, turkey, duck and opossum remains than those of farm-raised livestock.[194]

Although Georgetown slaves probably ate well-balanced meals consisting of assorted meats, fruits and vegetables, many health threats existed in the stagnant, foul-smelling rice fields of the Lowcountry. Germs grew in the inactive water, and rats, mosquitoes and other insects, snakes and other animals such as gators thrived. Dysentery and cholera were common.

Most of the medicines that Dr. Parker administered at Mansfield were to cure constipation. Drugs such as castor oil (thin yellow oil), jalap (a dried, leafy substance), calomel, rhubarb, Glauber's salt and magnesia are laxatives and purgatives. The use of laxatives might suggest that the diets of the slaves were high in starch and low in roughage, but this is probably not the case since slaves ate mostly locally grown foods. Antebellum doctors saw laxatives as cure-all miracle drugs, so the laxatives might have been used to cure an array of ailments.

Not only did slaves have freshly grown foods, but planters also made sure that their slaves had enough to eat to keep them healthy. One planter gave explicit instructions to his overseer concerning how to feed his slaves. He wrote, "Great care should be taken that the Negroes should never have less than their regular allowance: in all cases of doubt, it should be given in favor of the largest quantity. The measures should not be struck, but rather heaped up over. None but provisions of the best quality should be used."[195]Another one of Dr. Parker's contemporaries boasted of his ability to take care of his slaves in a letter to another planter. He claimed that his house servants "take no regular allowance but are dieted from my own table."[196] Concerning his field servants, the same planter wrote:

> *My habit has been for years to vary the weekly allowance of my people. As bread stuff, their diet consists of rice, corn, peas and potatoes with rations of molasses, salted fish, pork or bacon and fresh beef. The allowance for one week is meal (corn) 10 quarts, of rice or peas 8 quarts, and of sweet potatoes one bushel. Molasses is given throughout the year at proper intervals, salt fish only in the winter, pork or bacon and beef during the summer.*[197]

Besides administering laxatives, Dr. Parker also performed more serious procedures. The typical medical practices of the day included cuppings and bleedings and medicine including powders and powerful prescriptions. Among the medicinal supplies on hand at Mansfield, Parker lists numerous potent concoctions. He used ipecac to induce vomiting and snakeroot to clear the respiratory system. Of most interest, perhaps, is the use of laudanum (which contained cocaine and opium as well as alcohol) and paregoric (an opium-based drug) at Mansfield. These drugs were used to deaden pain but are considered to be highly addictive. Another compound, calomel, a colorless and tasteless concoction, was also used as an insecticide and fungicide, while turpentine was used in the treatment of lice. Magnesia was given and used as an antacid, and flaxseed was used for various derivatives.[198]

Being a trained physician, Dr. Parker certainly helped his slaves and family at Mansfield. But the doctor was not always in residence. We do not know what happened at Mansfield while Dr. Parker was away from the estate; however, we can look to his contemporary Plowden Weston's example of leaving strict rules for his overseers about how to deal with illnesses among the slaves when he was absent from his estate. Although not a definitive course of action, Weston's rules and procedures give valuable historical insight:

> *All sick persons are to stay in the hospital night and day, from the time they first complain to the time they are able to go to work again. The nurses are to be responsible for the sick not leaving the house, and for the cleanliness of the bedding, utensils, etc. The nurses are never to be allowed to give any medicine without the orders of the Overseer or Doctor. A woman, beside the plantation nurse, must be put to nurse all persons seriously ill. In all cases at all serious the Doctor is to be sent for, and his orders are to be strictly attended to; no alteration is to be made in the treatment he directs. Lying-in women are to be attended by the midwife as long as is necessary, and by a woman put to nurse them for a fortnight. They will remain at the Negro houses for 4 weeks, and then will work 2 weeks on the highland. In some cases, however, it is necessary to allow them to lie up longer. The health of many women has been entirely ruined by want of care in this particular. Women are sometimes in such a state as to render it unhealthy for them to work in water; the Overseer should take care of them at these times. The pregnant women are always to do some work up to the time of their confinement, if it is only walking into the field and staying there. If they are sick, they are to go to the hospital, and stay there until it is pretty certain their time is near. Nourishing food is to be provided for those who are getting better. The Overseer will keep an account of the articles he purchases for this purposes, during the Proprietor's absence, which he will settle for as soon as he returns.*
>
> *Bleeding is under all circumstances strictly prohibited, except by order of the Doctor. The Overseer is particularly warned not to give strong medicine, such as calomel, or tartar emetic; simple remedies such as flax-seed tea, mint water, No. 6, magnesia, etc., are sufficient for most cases, and do less harm. Strong medicines should be left to the Doctor; and since the Proprietor never grudges a Doctor's bill, however large, he has a right to expect that the Overseer shall always send for the Doctor when a serious case occurs. Dr.* [illegible] *is the Physician of the place. When he is absent, Dr.* [illegible]. *Great care must be taken to prevent persons from lying up when there is nothing or little the matter with them. Such*

> *must be turned out immediately; and those somewhat sick can do lighter work, which encourages industry. Nothing is so subversive of discipline, or so unjust, as to allow people to sham, for this causes the well-disposed to do the work of the lazy.*[199]

Concerning plantation nurses, Weston instructed the overseer to make sure that they followed his strict rules as well:

> *Nurses are to take care of the sick, and to be responsible for the fulfillment of the orders of the Overseer, or Doctor, (if he be in attendance.) The food of the sick will be under their charge. They are expected to keep the hospital floors, bedding, blankets, utensils, etc., in perfect cleanliness. Wood should be allowed them. Their assistants should be entirely under their control. When the Proprietor and Overseer are absent, and a serious case occurs, the nurse is to send for the Doctor.*[200]

Dr. James Ritchie Sparkman wrote a letter explaining how his slaves were treated when they were sick. Concerning medical attention, he wrote, "Valuable slaves, cherished family servants, like prominent and valuable citizens and heads of families, receive alike extra consideration and attention without reference or regard to cost." He finished his discourse on the matter by stating, "On every well-regulated rice plantation, the sick receive the first and, if necessary, the undivided attention of the overseer."[201]

Although records do not exist to prove that Dr. Parker wrote long and detailed rules and correspondences with his overseers or contemporaries concerning medical treatment of his slaves during his absence, we can gain valuable insight into how he promoted proper health among his slaves by reading his plantation journal. Among the supplies listed in his journal, Dr. Parker mentioned ordering a supply of fly hister. "Fly hister" was the term used to describe specific beetles that live in manure packs and in homes. Approximately 250 hister beetles can live per quart of manure, and 12 beetles can live comfortably within a square inch. When flies land on manure, the beetles consume the fly, which inevitably cuts down on disease transferred by flies.[202]

A more common method of keeping slaves healthy was giving them a half task on Saturday so that they had the second half of the day to clean their clothing, cabins, tinware, plates, bowls, eating utensils and other property. On Saturday afternoon, the planter or overseer inspected the slaves' cabins and belongings for cleanliness. As long as everything was clean, slaves received their weekly rations.[203]

Of course, even with fly hister, half days on Saturday to clean their belongings, nurses and even designated hospital buildings on some plantations, traveling doctors and even medically trained planters, slaves did get sick and needed medical attention. With planters being part-time residents due to health reasons that required them to leave the malaria-ridden rice swamps during the summer, and the draw of the social season in Charleston, slaves were forced to take matters of health into their own hands.

Like their ancestors in Africa, Lowcountry slaves used vegetation and other locally available products to create home remedies and herbal cures to their ailments. These voodoo witch doctors used various derivatives and created many concoctions that included, among other things, chimney soot, cockroaches, twigs, tree sap, herbs and flowers, snails and even animal droppings.

### Lowcountry Concoctions[204]

| **Symptom** | **Product** | **Method** |
|---|---|---|
| epilepsy ("the fits") | cockroach | crushed and boiled in a tea and drank |
| whooping cough | goat or sheep droppings | boiled to a tea and drank |
| warts | snail | crushed and applied to wart to dry it out |
| bleeding | spider webs | applied to cut |
| pneumonia | hogs hoofs | cleaned, parched, pounded, boiled and strained and drank in a liquid form |
| hemorrhages | chimney soot in conjunction with spider webs | rubbed in wounds |
| labor pains | chimney soot | drank in a tea to hasten labor |

| Symptom | Product | Method |
|---|---|---|
| diarrhea or indigestion | chimney soot and hickory ashes | half gallon of boiling water over one quart of hickory ashes and one teacup of clean soot; after cooled, the concoction was consumed three times per day |
| rheumatism | red clay | wrap ankles |
| headache | canna flower | bound to the head |
| fever | canna leaf | bound to the back of the neck to bring down fever |
| fever and chills | corn husks and red pepper | boiled together and drank as a tea |
| flatulency or ulcers | calamus, also known as flag root | boiled to a tea |
| coughs, colds and constipation | star grass | infused with vinegar |
| high blood pressure and fungus | wild leek | cooked in soups |
| snake bite, rheumatic and coetaneous ailments | devil's stick or white root | boiled in tea to increase appetite and overcome fever |
| throat irritation and whooping cough | pincushion flower | leafy branches are boiled with sea myrtles and then gargled; the inner root treats palsy and venereal and skin diseases and halts hemorrhages |
| ground itch/athlete's foot | sweet pepper bush | the root is boiled, Epson salt added to the concoction and feet soaked in it |

| Symptom | Product | Method |
|---|---|---|
| thrush (locally known as thrash) in infants | persimmon, also known as winter plum | a slice of green persimmon is rubbed on the baby's tongue and mouth to remove thrush; it is also rubbed on the gums of a teething child to relieve the pain |
| bleeding gums, stuffy nose, diarrhea or swelling | witch hazel | witch hazel clears nasal passages and, when rubbed on bleeding gums, can stop bleeding; the bark is taken as a remedy for diarrhea and also reduces swelling |
| fever | Spanish daisy | drank in a tea |
| coughs and colds | American holly | the root can be boiled to a tea and taken for colds. For bad colds or fever, the root was boiled with life everlasting, pine tops and lemon juice |
| kidney stones, smallpox, diabetes | Yaupon leaf | drank in a tea |
| stomach trouble and diarrhea | honeysuckle root | boiled and then taken for stomach trouble and diarrhea; also, bathing in honeysuckle is believed to promote vitality and longevity |

| Symptom | Product | Method |
|---|---|---|
| worms | chinaberry tree leaves | chopped and boiled; this decoction was regularly given to children to expel worms |
| aching back, jaundice, scrofula, diarrhea, dysentery and uterine hemorrhages | wax myrtle leaves | steeped for tea and used as a remedy for aching backs because it clears out the kidneys; the leaves can be boiled with twigs as a remedy for diarrhea, severe colds and fever; it has been used to treat jaundice, scrofula, diarrhea, dysentery and uterine hemorrhages; it has also been taken as snuff to treated nasal polyps; the powdered root is used for sore throat and mouth, and used to wash ulcers and tender and bleeding gums |
| diarrhea and genital itch | water lilly | tea for diarrhea and bath for genital itch |
| gonorrhea, gleets, leucorrhea, chronic diseases of the urinary passages in chronic catarrhal affections, chronic rheumatism, piles, chronic inflammation of ulceration of the bowels | longleaf pine | boiled to a tea |

| Symptom | Product | Method |
| --- | --- | --- |
| diarrhea, dysentery, gonorrhea, hemorrhoids or syphilis | sumac root | when peeled and chewed, sumac root provided a remedy for diarrhea and dysentery; the leaf was used for bathing painful places on the body, while the root was boiled with blackberry root and pine tops and taken for venereal diseases, especially gonorrhea; sumac was used for diarrhea, dysentery and hectic fever and was frequently used in a concoction mixed with barks of slippery elm and white pine for relief from syphilis |
| pain during childbirth, low milk flow, high blood pressure, hemorrhoids and neck pain | Spanish moss | when green, the moss was boiled in tea to help ease delivery and increase the flow of the mother's milk; the green moss was also put in shoes to remedy blood pressure; the crushed moss was used to relieve hemorrhoids and, when green, to help with neck pain |

| Symptom | Product | Method |
|---|---|---|
| toothache, fever, ulcers, chronic diarrhea and dysentery | swamp willow | swamp willow has been used as a substitute for quinine in treatment of fever; tt was also used as a stringent in fighting chronic diarrhea and dysentery |
| lack of potency in men and to regulate menstruation in women | ironweed | the root of ironweed was taken by men to increase their potency and by children to relieve fever; the root was chewed to relieve toothache, and a decoction of the plant was taken for indigestion and to regulate menstruation |
| unwanted pregnancy | cotton | a root tea made from cotton was taken as a contraceptive on the plantations and also used to induce abortion |

There is no way to determine which of these concoctions, or perhaps others that were not mentioned, were commonly used at Mansfield Plantation. We do know, however, that herbal medicines and home remedies were an important part of the slave experience in the South Carolina Lowcountry.

# *Chapter 5*

# Labor and Social Status on Georgetown Rice Plantations

*The distinctions separating the social classes are false; in the last analysis they rest on force.*
*—Albert Einstein*

Some planters constructed fine examples of Colonial, Provincial, Republican, Georgian and Greek Revival manor homes along the banks of the South Carolina Lowcountry rivers. Flowering magnolia, camellia trees, azalea gardens, flowering vines, labyrinthine hedges of roses and long avenues of live oaks complemented the landscaping leading to and around the manor houses—the seats of the rice growers' personal empires.

Other planters lived more humble lifestyles in comparatively smaller homes built on simple brick foundations. Some plantations did not have a manor house at all. In many cases, a planter would own several plantations but build only one house for his family. On the plantations where no planter lived, an overseer lived in a small house and managed the production of rice for the absent landowner.[205]

Each plantation in Georgetown was its own self-sufficient entity, capable of producing everything that a community needed to survive. Although rice was the most abundantly grown crop on the Georgetown plantations, it was certainly not the only one. Other commonly grown crops included corn, potatoes, squash, peas, beans, tomatoes, gourds, figs, wheat, sweet potatoes, melons, apples and peaches. Most plantations also raised pigs, cows and chickens for consumption on the estates and sometimes for sale. The inhabitants of the Georgetown plantations, both master and slave, also depended on wild game such as deer, raccoon, opossum, turtles, alligator, fish and wild duck.[206]

Each owner designed his own home. After procuring Mansfield, Dr. Francis S. Parker added a wing to the preexisting manor house to accommodate his family's growing needs. He also constructed a new slave village with a chapel, schoolhouse and freestanding kitchen.[207]

Each planter made many decisions when he designed a plantation settlement besides the architecture of the manor house. He considered the number, type and proximity of the slave dwellings, overseer's quarters, driver's quarters, livestock stables, gardens, kitchens, winnowing houses, rice-pounding mills, threshing mills, toolsheds and any other permanent structures. Most of those decisions were based on the topography of the land and proximity to the river, not aesthetic reasons.[208]

The planter also decided on the treatment and lifestyle of his slaves. Although giving their slaves alcohol and guns for hunting and educating house servants were illegal actions, planters used their own discretion in matters of naming, feeding, housing, clothing, administering religion, arming and burying slaves.[209]

All the plantations retained the same hierarchy and social structure. At the top of the social structure was the planter and his family. Of course, males dominated southern society and were viewed as cavaliers and gentlemen, strong and masterful. Women were dominated by their male counterparts, and many, including Anna Maria Taylor Lance and her two daughters, Mary Lance Parker and Ester Jane Lance Read, were burdened with numerous pregnancies. However, women had very specific roles and dominated the manor house as their domain.

During their childhood, girls were nursed by elder slave women and later learned socialization through play with the slaves of their community. Girls began their training as young ladies by traveling with their mother to visit family and friends and attending church with their nurse. Mothers were also responsible for teaching their daughters manners. Soon, mothers taught their daughters to behave, to read and write, to sew for the planting family (a seemingly endless task of repairing clothing as well as sewing new outfits) and gardening. Young ladies were also responsible for keeping their rooms and clothing clean and arranging flowers and putting up preserves.[210]

Directly under the planting family was the overseer, an agricultural specialist who was usually the only other member of the Caucasian race living on the plantation. Overseers were generally literate and had great interest in upholding the plantation system since most aspired to become planters themselves. They were in charge of discipline and everyday treatment and health of slaves on the plantations, as well as crop production and the general upkeep of the estate. Overseers earned a good living compared to the poor white community, generally referred to as the yeomanry. Some owned property, including slaves. Some overseers abused their authority and, in turn, abused slaves.

On at least one occasion, a planter fired his overseer for abusing slaves left in his care. Fredrika Bremer, a Swedish visitor to the United States in 1849, recalled the event: "Extreme cruelty was practiced against the slaves on a plantation in the neighborhood by an overseer, during the prolonged absence in England of the owner of the plantation. The planters in the area united, wrote to him, told him and requested that the overseer should be removed. And this was done."[211]

Joseph Allston recorded another incident of an overseer's excessive punishments to slaves: "Your Jack…was most dreadfully beat on Pee Dee by Swinton, General Carr's overseer, so much that he has not been able to work since…If the punishment was as severe as represented to me, I think Swinton (the overseer) ought to be prosecuted as nothing could justify it."[212]

The master of Hagley Plantation on the Waccamaw River gives us a glimpse inside the world of planter-overseer relationships and the responsibilities of the overseer in his detailed description of what he expected from his overseer and how the agricultural specialist would be compensated for his services:

> *The Proprietor wishes particularly to impress on the overseer the criterions by which he will judge of his usefulness and capacity. First—by the general well-being of all the negroes; their cleanly appearance, respectful manners, active and vigorous obedience; their completion of their tasks well and early; the small amount of punishment; the excess of births over deaths; the small number of persons in the hospital, and the health of the children. Secondly—the condition and fatness of cattle and mules; the good repair of all the fences and buildings, harness, boats, flats, and ploughs; more particularly the good order of the banks and trunks, and the freedom*

> *of the fields from grass and volunteer. Thirdly—the amount and quality of the rice and provision crops.*[213]

Concerning punishments for slaves, one planter wrote that he hoped the overseer "will remember that a system of strict justice is necessary to good management. No person should ever be allowed to break a law without being punished, or any person punished who has not broken a well-known law. Every person should be made perfectly to understand what they are punished for, and should be made to perceive that they are not punished in anger, or through caprice." Concerning specific punishments, he wrote, "It is desirable to allow 24 hours to elapse between the discovery of the offence, and the punishment. No punishment is to exceed 15 lashes...Confinement (not in the stocks) is to be preferred to whipping, but the stoppage of Saturday's allowance, and doing whole task on Saturday, will suffice to prevent ordinary offences. Special care must be taken to prevent any indecency in punishing women."[214]

In 1844, in the inaugural edition of the *Carolina Planter*, a contributing author explained punishment on his estate as a detailed example for overseers to follow: "Negroes who will not do their work, like boys who will not get their lessons, must sometimes be flogged; but a few stripes are all that is necessary…Fifteen or twenty lashes lightly inflicted are as much as should ever be given. For serious offenses, other punishments, such as solitary confinement, etc. should be resorted to."[215]

In January 1847, Dr. Parker noted in his journal that he paid his overseer, J.B. Harrell, the sum of $300.00 per year as a base salary but also promised to compensate Harrell a percentage of the crop yield. Parker agreed to pay Harrell $0.05 for every bushel of rice over seven thousand sent to market and $50.00 extra if sufficient provisions were made for the plantation so that none had be purchased.[216]

Parker's overseers were doing well in the decade before South Carolina declared independence. By 1850, Parker hired Francis W. Mummerlyn to oversee Mansfield. Parker agreed to pay him $300.00 per year and $0.05 for every bushel of rice over six thousand sent to market, as well as a $50.00 stipend for successfully running Mansfield. By the early 1850s, Mummerlyn was paid $400.00 per year, plus the stipends he had received in the past, to oversee all of Parker's rice interests. Sub-overseers were hired to run the operations on the individual Parker estates, while Mummerlyn became increasingly more valuable to Parker. By 1860, the year before the war started, Mummerlyn's salary was $1,400.[217]

In 1850, Francis W. Mummerlyn, the forty-one-year-old overseer, owned $10,000 worth of property and presumably lived rent free at Mansfield Plantation. Mummerlyn, along with his wife, Sandra (thirty-four); their four sons, ranging in age from twelve to one; W.S. Mummerlyn (perhaps a son from an earlier marriage or a relative); a twenty-two-year-old clerk; and Thomas Collins, an eleven-year-old boy, lived on the Parker property. In 1850, the overseer owned three slaves—two males aged eleven and sixteen and a sixty-year-old female. By 1860, Francis W. Mummerlyn had increased the size of his own personal gang of slaves to an eight-member family of slaves. The father was thirty-six and the mother thirty. They had three girls (ages fourteen, twelve and eight) and three boys (ages ten, six and two).[218]

One of Parker's sub-overseers, J.J. Cox, owned three female slaves in 1860, including a thirty-two-year-old mulatto woman and two black slaves aged twenty and eighteen. Without records, it is hard to determine exactly what these slaves did; however, they probably worked for Dr. Parker and earned their master additional income for their services.[219]

A statistical analysis of the eighty-three overseers in the Georgetown District in 1860 reveals that their average age was 30.4 years and that only 12 percent were over 40.0 years old. A total of 58 percent were married, none were illiterate, 30 percent owned real property, 18 percent owned substantial property and 22 percent owned slaves (average number of slaves owned was 8.6).[220]

Under the overseer was the driver. Drivers were slaves who had high stature in the slave communities and held the authority, invested by the planter, to be the boss over the field hands. It was the driver's job to make sure that the slaves worked and obeyed the overseer. In essence, the driver was a mediator between the slaves and the overseer, who was essentially the mediator between the field slaves and the planter.

Drivers were "under the overseer, to maintain discipline and order on the place. They are to be responsible for the quiet of the Negro-houses, for the proper performance of tasks, for bringing out the people early in the morning, and generally for the immediate inspection of such things as the overseer only generally superintends."[221]

Approximately two and a half hours before sunrise, the driver instructed the night watchman to ring the plantation bell to wake slaves for work. The slaves then reported to the driver, who not only rode the flat-bottomed boats with the field hands to the rice fields each morning but also clearly pointed out to each field hand his or her daily task. The driver, with the aid of a helper, clearly marked all of the fields to be worked that day before the workers awakened. He also inspected each field hand's work before dismissing him for the day.[222]

The driver not only had social status over the other slaves on the plantation but also had better material possessions. At Mansfield, the driver's cabin was at the end of the slave street and closest to the overseer's cabin. In Francis S. Parker's Mansfield Plantation journal, Parker always noted that his drivers, Dick, Zack and later Colby, received extra clothing at allowance time and a blanket each year instead of the biannual allotment and a "great coat." At Parker's Willowbank Plantation, drivers Pomphey and later Cyrus received the same benefits.[223]

In selection of drivers, planters looked to size and strength—most drivers were reportedly tall and strong as well as of good judgment and intelligence. A driver's authority was not limited to the field work but also to the village. Drivers were often policemen and even magistrates over the slave community.[224]

On some plantations, the ascension to the position of driver was ritualistic and cause for a great public ceremony. In some instances, the appointment to the position of driver was confirmed by a bishop. This ceremony showed the other slaves that God and his master trusted a certain member of their community to lead the others and expected the others to follow his instructions. The ceremony promoted a mystical and religious aura to the job and to the power and status of the position.[225]

Besides the driver and house servants, it is difficult to see social status by simply looking at the names of Parker's gang in his ledger. Parker simply lists clothing allotments and supplies dispersed twice per year. For example, on November 29, 1846, Parker handed out winter supplies to, what he referred to as, "the people." Dick, the driver of the others, received 8 yards of clothing (probably enough material to make five or six outfits), a blanket and a "great coat," while the typical allotment for an adult was 6 yards of clothing (probably enough material to make four or five outfits) and less than that for a child. The entire gang numbered sixty-eight, of whom seven were listed in a separate list as house servants. In 1848, Parker dispersed another typical allotment of clothing and supplies to his eighty-three slaves at Mansfield: 400 yards of plains, three dozen caps, three dozen handkerchiefs, five pounds of thread, one hundred needles and fifty-eight pairs of brogans (shoes). In 1850, Parker handed out clothing supplies to ninety-three slaves at Mansfield, including 425 yards of plains, three dozen handkerchiefs, three dozen caps, four pounds of thread, three gross of buttons, sixteen blankets, sixty pairs of brogans and one hundred needles. Glauber's salt, according to Dr. Parker's plantation journal, was used in the dyeing process of clothing.[226] Without great discrepancy in typical clothing material and supplies allotments, we must look past material possessions in order to find social status among Parker's slaves.

Since Mansfield had a steam-powered threshing mill during the antebellum era, Dr. Parker must have had a slave whom he trusted as the watchman or engineer, even though his records do not note anyone as serving in that capacity. The watchman was an intendant or steward. He carried keys to all of the barns, including those used for storing produce, provisions and tools and materials; the smokehouse; and the carpenter and smith shops. He was responsible for the maintenance of the mill and usually had his own house. The watchman was usually the wealthiest of all slaves (in terms of material possessions) on the estate and usually received valuable gifts from the planter to decorate his cabin and show his prominence on the estate.[227]

Slaves were generally broken into three categories, and their status was determined by their occupations. The three groups of slaves were house servants, skilled artisans and agricultural laborers, who were at the bottom of the social hierarchy. The latter constituted the largest group of slaves and had the least amount of contact with their master.[228]

Positions held by domestic servants included but were not limited to cooks, servers, herbal healers, maids, butlers, carriage drivers, laundresses, weavers and seamstresses. These slaves displayed proper manners, were better dressed than the others and had the closest access to their master and his material possessions. In 1937, a former field hand explained in a Works Progress Administration interview, "Just the house servant get Marse Josh and Miss Bess ear."[229]

As a result of their contact with their master and their training, house slaves were generally considered an upper class of slaves. They usually slept in either the manor house, a standalone kitchen or in domiciles constructed close to the manor house. Generally, these servants had little contact with the other slaves, and they traveled with the planting family to the beach during the summer months and to Charleston during the social season.

Domestic servants made up about 10 percent of Dr. Francis S. Parker's total gang. From 1849 to 1853, Parker staffed Mansfield with six house servants and kept five additional house servants in Charleston. By 1858, Parker's staff had increased to ten at Mansfield and seven in Charleston.[230]

House servants were difficult to purchase, so most planters selected young girls from their community and trained them to be house servants. Young girls spent their time in play with the planter's daughters until age seven or eight, at which point they began their training by caring for younger slave children. Soon thereafter, they were given an opportunity to audition for house servitude. If they proved to show potential, they were fully trained in homemaking arts.[231]

Like the young girls, young slave boys played with the planter's sons and learned how to hunt and fish. At seven or eight years of age, boys began to work at small farm tasks such as working with livestock. Some were selected and given an opportunity to try out for work as a house servant.[232]

Typical livery colors for house servants were green, red or gray; however, planters clothed their servants as they wished. In this paternalistic world, planters knew that the appearance of their people was a direct reflection of their ability to provide. Dr. Francis S. Parker did not leave records addressing the uniform of his domestics, but knowing Parker's status among rice planters, we can be sure that he dressed them similar to the way that his contemporaries dressed their domestics.

One planter on the northern end of the Waccamaw River dressed his male domestics in fine dark green coats of broadcloth, green plush trousers and vests faced with red and trimmed in silver braids.[233] A contemporary on the Pee Dee River proudly boasted that his male domestics were "supplied without limit to ensure a genteel and comfortable appearance."[234] Planters sometimes clothed their family in the same material that they used to make suits for their servants. One planter asked his wife to tell his tailor, "If there is enough of it left (referring to a particular type of tweed), I would like a coat of it for myself." The family outfitted their female maidservants in cotton plaids and male domestics in gray, mixed-cloth coats and tweed trousers.[235]

Sometimes trouble existed if there were too many house servants at an estate. One plantation mistress had difficulty managing large domestic staffs. She wrote to her husband complaining that she had too much help:

> *There are too many servants; I do not know what to do with them. There is Mary, the cook; Milly, the laundress; Carolina, the house-maid; Cinda, the seamstress; Peter, the butler; Andrew, the second dining-room man; Aleck, the coachman; and Moses, the gardener. And George, the scullion, and the boy in the yard besides! I cannot find work*

> *for them! After breakfast when they line up and ask, "Miss, wha' yu' want me fu' do to-day?" I feel like running away. Please send some of them away, for Lavinia is capable of doing the work of two of them. Please send them away, half of them, at least.*

The husband responded by explaining that it would be cruel to send house servants to work in the fields because they were not trained for that sort of labor.[236]

Skilled artisans performed various tasks on the plantations and in towns and cities. They filled the necessary positions of carpenters, blacksmiths, brick masons, coopers, potters, silversmiths, porters, boat pilots, engineers, mill operators, tailors, musicians, butchers, barbers, shopkeepers and vendors. Some of these skilled craftsmen were very talented.

One carpenter slave was sent to England to be trained. Trained carpenters were expected to train a crew of their own on the plantation to assist them in building construction. The crews were responsible for constructing not only the main house and slave quarters but also kitchens, barns, truck docks, floodgates, mills and every other structure on the estate.[237]

Concerning task assignments, Charles E. Rowland, Esquire, a member of the South Carolina Agricultural Society, published an article entitled "Highly Valuable Information for Rice Planters" in Georgetown's *Pee Dee Times* on January 31, 1855. Rowland gave rice planters valuable insight into what should be expected from skilled artisans as a complete task:

> *Making rice barrels—Three barrels a day is the task for a cooper, and in making half barrels, four a day. The length of the stave* [should be] *three feet, two inches; and two feet across the head. Such a barrel will hold ten bushels of clean rice.*
>
> *Hoop pole—Whereas hoop pole are plentiful, a negro can with ease cut one hundred, and bring them home, where the distance is not too great.*
>
> *Splitting shingles—In splitting shingles 22 inches long, four hands are sent out: two to cut the tree and cross cut to the length, one to bolt, and the fourth to split one thousand. The next day, a negro will draw 500 per day as his task. In splitting for negro houses, he will split 700 broad, heavy cypress shingles two and a half feet in length. In drawing 400 is the day's work.*

> *Squaring timber—A negro carpenter can cut down the tree and square one hundred feet per day with ease.*
>
> *Post and rail fences—Four negroes can put up 35 to 40 panels of post and rail fence per day; dig the holes between two and three feet deep and put down the posts properly, rammed, at a distance of nine feet from each other. The rails are to be ten feet long, to allow a good lap or hold in the mortise. A negro carpenter can make 60 mortises per day, in the post and rail fence.*
>
> *Splitting rails—One hundred rails, of 12 feet long, and heavy, is the day's work of an able bodied negro man.*
>
> *Cutting wood—A negro man can cut for his task, one cord of oak wood, 4 feet long. The cord when piled and well filled is to be 8 feet long, 4 feet high.*[238]

Agricultural laborers were generally considered unskilled laborers. Tasks performed by agricultural laborers (field hands), besides planting and harvesting crops, included the roles of plowmen and ditchers, as well as axe men, gardeners, hunters, stable hands, trappers and jockeys. Parker's journal provides numerous examples of planting crops at Mansfield. One example covers the 1847 planting season. In mid-February of that year, Parker's overseer, J.B. Harrell, instructed the gang to plant fifteen acres of oats. During the last week in March, the gang planted rice. They planted corn in early April. In May, slaves planted a second rice crop.

Rowland's article outlined not only work for skilled artisans but also what planters and overseers should expect as typical daily tasks and labor quotas for field hands. While watching field workers perform their tasks, one observer noted that most slaves completed their tasks by two o'clock, with some completing their tasks before one o'clock. He claimed that under the task system, most slaves worked rapidly and well and that the tasks would not be considered to be excessively hard work by a northern laborer.[239]

> *Ditching—In ditching, much depends upon the quality of the ground. In old rice fields, free from roots and stumps, the task for an able bodied negro man is 600 cubic feet; but he cannot do as much*

*in canals. Ditching is much retarded by gravel and iron mould as well as by roots and stumps. When these obstacles occur, no regular task can be assigned. If he excavates three or four hundred feet, he will do well.*

*Clearing large cypress trees—Where large cypress stumps are met with, the best plan is to select Negro men who are best with cutting with an axe. A very large cypress stump will take one to two days work. In ditching, it is advisable to work in gangs of six or eight, in a given distance, consisting of men and women. Three men can handle the spade, while the women haul out the excavated earth.*

*Making bank—In making bank, it will take one or two less in number than the same length in ditching. Where the soil is stiff clay, and the distance of the margin 20 feet, it will take full complement to have the bank well made and trimmed properly.*

*Turning up land with a hoe—A negro man or woman, can in light rice land of a deep mould, turn up one quarter of an acre and do it well; but in stiff tenacious clay, it will require a third more labor.*

*Listing ground—In listing corn, cotton, or potatoes land, half an acre is the task for each hand; but in old pasture ground which is always stiff and bound with roots of grass, weeds & etc, a quarter of an acre is hard work. Where it is of a light and sandy nature, half an acre can be accomplished in listing.*

*Bedding up—The task in ridging or bedding up, is generally a quarter and a half, but as expressed above in pastureland, where the texture is close, he will do well if he does one quarter. In ridging potato heads, two negroes are always employed in one quarter of an acre.*

*Trenching rice—Each negro man can trench half an acre of rice land, containing 70 or 80 rows in one quarter, and each negro woman covers what the man trenches. In planting rice, the more compact your gang is the better. A gang consisting of forty workers may be divided into two parts, allotting to each two sowers, nine trenchers, and nine covers, in which case each gang would plant five acres.*

> *Hoeing corn, cotton potatoes—In hoeing corn, cotton and potatoes, the usual task is half an acre; but if neglected in the first hoeing, the negro will find it difficult to overcome the grass in the succeeding ones, particularly cotton and potatoes, which must be kept very clean.*
>
> *Hoeing rice—In hoeing rice, in its early stage, much depends upon the judicious application of water, to destroy grass; improperly applied, it encourages it, and increases the labor of the negro. He can always get through half an acre with ease during the season. In the first hoeing on clay land, less than three negroes to the acre cannot perform their work as it should be done and great care taken to remove the clods from young rice, which would otherwise be much injured. In the third and last hoeing, every spear of grass ought to be pulled up by the roots to allow the shoots to come forth freely.*
>
> *Threshing rice—Six hundred sheaves of rice is the general task for a negro man and five hundred for a negro woman, which can be done with ease; after which, the straw should be well shaken and examined by the driver before it is carried to the general heap.*[240]

An antebellum planter on the Pee Dee River explained to a friend how he assigned tasks and worked his field hands:

> *The ordinary plantation task is easily accomplished, during the winter months, in 8 to 9 hours and in summer my people seldom exceed 10 hours labor per day. Whenever the daily task is finished the balance of the day is appropriated to their own purpose. In severe freezing weather no task is exacted, and such work is selected as can be done with least exposure. During heavy rains and in thunder showers, my people are always dismissed and allowed to go home. The task is allotted to each slave in proportion to his age and physical ability. Thus they are considered ¼, ½, ¾, or full task hands. Allowance is invariably made for women so soon as they report themselves pregnant they being appointed to such light work as will insure a proper consideration for the offspring. No woman is called out to work after her confinement, until the lapse of 30 days, and for the first fortnight thereafter her duties are selected on the upland, or in the cultivation of the provision crops, and she is not sent with the gang on the low damp tide lands. Men and women are all engaged together in the planting, cultivation and harvesting of the crop, but in*

> *the preparation of the rice lands, as ditching, embanking, etc. The men alone are engaged with the spade. It is customary (and never objected to) for the more active and industrious hands to assist those who are slower and more tardy in finishing their daily task.*

In 1858, another local rice planter explained the task system on his estate in one of the leading agricultural journals of the era, *De Bow's Review*:

> *Each full hand (grown person) is required to turn or dig up one quarter of an acre of swamp land (rice land) per day. In cutting ditches, the task is 600 feet...In listing land each Negro will do half an acre; in bedding land three eighths of an acre; in trenching land for rice each man will trench three-fourths of an acre which will contain 180 rows...*[241]

Although merely suggestions, the written records left by contemporaries provide a window into understanding the status and workloads of slaves. However, individual planters reserved the right to take care of and manage their slaves in the fashion they wished. Besides freewill, another determiner of the treatment of slaves was based on the circumstances of how masters could afford to provide for their slaves with ration allotments and task expectations.

## *Chapter 6*

# Capitalism, Material Wealth and Christianity on Georgetown Rice Plantations

*The most important single central fact about a free market is that no exchange takes place unless both parties benefit.*
*—Milton Friedman*

Many historians have attempted to understand and explain the life ways of slaves in the antebellum South, but far too often they lump slaves and the institution of slavery into one broad statement—working sunup to sundown without any freedom or possession. These assumptions are very far from the truth concerning rice plantations.

One topic that most tend to ignore is the presence of money in the slave communities. One rice planter, Daniel Elliot Huger Smith, explained the informal economy of slaves. He wrote, "A good deal of money in a small way circulated among the Negroes, and the keepers of small grocery shops made a good profit out of this trade. These were chiefly Germans, who in many cases were bold breakers of the law against selling liquor to the blacks…These shops were always at corners and were called 'Dutchmen corner shops.' This aspersion on the Hollander was very unjust, for of them there were few and they were merchants of good standing and repute."[242]

Material possessions and leisure affected the quality of slaves' lives and created a gauge of wealth among slave communities. Exactly how influential and useful money was to bondsmen, or how widespread the use of manufactured goods, cannot fully be determined, but the mere fact that some slaves had money and finished goods changes the popular conception of American slavery.

On some plantations, planters permitted their slaves to sell goods that they produced to the yeoman farmers of the pine forests for either money or trade items. One former slave, Margaret Bryant, recalled one such incident of how the slave economy operated: "Po-buckra come there and buys cloth from ma. Buy three and four yard. Ma sell that, have to weave day and night to make up that cloth to please oberseer. Come big day time. Little chillum, 'Whey, mama?' Tell 'em Ma to the weaving house. Don't have money fuh pay. Bring hog and such like as to pay."[243]

Given the fact that all slaves had different skills, contacts and jobs to perform, it is not hard to understand that not all slaves had the same opportunity to earn money. Another former slave recalled his father's method of raising money:

> *My father love he liquor. That take money. He ain't have money but he have the rice barn key and rice been money. So my father gone in woods (he have a head my father), take an old stump, have 'em hollow out. Now he (the stump) same as mortar to the barn key. And my father keep a pestle hide handy. Hide two pestle. Them pestle make out heart pine. When that pestle been miss (missed) I wuzn't know nothing. The way I know my age, when the slavery time war come I been old enough to go in the woods with my father and hold a lightard (lightwood) torch for him to see to pestle off that golden rice he been tote out the barn and hide. That rice he been take to town Sat'd'y (Saturday) when Colonel and my father go to get provision like sugar, coffee, pepper and salt. With the money he get when he sell that rice he buy liquor. He been hide that sack o' rice fore day clean (daylight) in the prow of the boat and cover with a thing like an old coat.*[244]

Recorded interviews with ex-slaves give valuable insight into Georgetown's slave economy, but some planter's records are more complete. In 1858, one planter wrote to another describing the informal slave economy on his plantations. He wrote, "As an illustration of the indulgences which my own people enjoy, I have during the past year kept an item of their perquisites from the sale (to me) of eggs, poultry, provisions saved from their allowance, and the raising of hogs, and it amounts in money to upwards of $130.00 and in sugar, molasses, flour, coffee, handkerchiefs, aprons, homespun, and calico, pavilion gauze (mosquito nets), tin buckets, hats, pocket knives, sieves, etc. to the amount of $110.00 more. One man received $25.00 and another $27.00 for hogs of their own raising and I had the satisfaction of seeing most of these amounts spent in comforts and presents to their families."[245] He continued by saying:

> *The question has frequently presented itself whether or not this habit of buying for my people articles which are ordinarily only to be obtained through shop keepers, and retailing to them in exchange for their poultry etc. etc. might not lead to some abuse and cause them to pilfer or speculate upon each other. Limiting this rule strictly to my own people I have found it to work well. They frequently ask me to become treasurer of their little fund. I have become satisfied that they get better bargains than can be made by themselves with the shop keepers. I hold now in this way upwards of $70.00, which will be called for in dimes and quarters, through the current year. As a system this cannot be carried out in detail generally, for only those who are permanent residents, can very well undertake it. But the custom prevails with most of my neighbors of supplying all reasonable demands by exchange as indicated, whilst some prefer paying in money and allowing them to trade for themselves. The point established is that by reasonable industry and ordinary providence, our people all have it in their power to add materially to their comforts and indulgences, and that their owners very wisely and humanely offer every encouragement to this effort.*[246]

Many planters commonly purchased assorted vegetables, livestock, eggs, honey and firewood from their slaves. Some planters even paid their servants for performing more work than was generally expected of them. Robert F.W. Allston paid his coopers twenty-five cents for each barrel they produced over their daily quota of three. He also purchased pumpkins, firewood, chickens, honey and hogs from his slaves. By 1859, Allston gave up raising hogs altogether and paid his slaves five dollars per hundred pounds to raise them.[247]

Some planters paid slaves for working overtime and on days reserved for their leisure. One planter wrote careful instructions to his overseers regarding the management of his slaves: "No work of any sort or kind is to be done by Negroes on Good Friday, or Christmas Day, or on any Sunday, except going for a doctor, or nursing sick persons; any work of this kind done on any of those days is to be reported to the proprietor, who will pay for it…A task is as much as the meanest [healthiest] full hand can do in nine hours, working industriously. The driver is each morning to point out to each hand their task, and this task is never to be increased, and no work is to be done over task except under the most urgent necessity; which overwork is to be reported to the proprietor, who will pay for it."[248]

On at least one occasion, slaves were rewarded money for performing dangerous work. During the summer of 1854, while the planters were at their summer homes, three field hands put out a fire that undoubtedly would have consumed the planter's clubhouse at Guendalos Plantation on the Pee Dee. The Winyah and All Saints Agricultural Society, which held its meetings at the clubhouse, voted to pay the firefighting slaves a reward sum of $5 each for dousing the flame. This reward might seem small, but taking into consideration that the building was eleven years old and the construction cost was only $550, the slaves received a fair reward.[249]

Some slaves were fortunate enough to have marketable skills. In many cases, slaveholders allowed those slaves to live in town and hire out their services; of course, the slaveholder received about half. Daniel Elliot Huger Smith once wrote, "Many Negroes, especially the mechanics, were allowed to 'work out' or to 'hire their own time.' For these the city authorities issued a license." Hired-out servants usually worked contracts that ran from January through Christmas.[250]

Among the more popular occupations of these "hired-out" slaves were coachmen, carpenters, coopers, barbers and blacksmiths among men. Women "hired out" as cooks, seamstresses and maids.

On at least one rice plantation, the planter had a store that was managed by a slave woman. The planter made it a rule on his plantation that his store bought everything that his slaves wanted to sell. The sellers were rewarded with either cash or credits for their items. The planter bought large quantities of items that the slaves wanted most, including ammunition and firearms, and passed his wholesale prices on to his slaves. Slaves became the planter's greatest creditors, as each one had his or her own account with him.[251]

Another symbol of slaves' quality of life can be seen in the holidays and freedoms that they had on plantations. Of course, all planters were free to work their slaves as they saw fit, but all masters gave slaves some holidays. Common practices of celebrating on plantations included no work of any sort or kind done by slaves on Good Friday, Christmas Day or any Sunday. Other holidays included the two days following Christmas

Day and the first Saturdays after finishing threshing, planting, hoeing and harvesting.[252]

Slaves were normally given the opportunity to leave the plantation to visit friends on nearby plantations as long as they were home by nightfall or, in the case of a married couple, on time for work in the morning. No one was to be absent from the plantation without a list ticket, which was usually given provided that the slave behaved. All persons coming from the other places had to show their tickets to the overseer, who would sign his name on the back. Those who went off the plantation would bring back their signed tickets. Overseers often traveled around the plantation village at night and took roll at each house to be certain that all slaves were home.[253]

In many cases, slave children enjoyed their lives in the country settings of the rice plantations. On some plantations, slaves did not have adult labor responsibilities until they reached the age of twelve or even older. To fill their days, many children listened to stories, including superstitions, gossip and folk tales told by superannuated (retired) slaves. Additionally, children visited friends and relatives on nearby plantations; helped care for younger children; frightened birds from agricultural fields; cared for livestock; tended to family garden plots; role played, hunted and fished; made baskets, jewelry, corn husk dolls and decorative pots; and played games. Some of the games that children played were marbles, hop scotch, jump rope, ball games, hide-and-seek, horseshoes and cards games, as well as games to test athletic abilities such as swimming, lifting, jumping and throwing contests, wrestling, boxing matches and footraces.[254]

The cooking of a hog provided entertainment and food for many people. Sometimes, slaves and masters would spend time cooking hogs together. The hog provided entertainment as well sustenance that could last for a week or more. Ribs, bacon and other meats would quickly be consumed, but lard, fat and cracklin' from the skin could be added to flavor basic one-pot dishes.[255]

A generation after the American Revolution, a social movement called the Second Great Awakening brought forth a period of great religious revival in the United States that extended into the antebellum period. The movement was responsible for great widespread Christian evangelism that is usually associated with the northern and middle states and, to a much lesser degree, the southern backcountry. It was responsible for many Christian conversions and the expansion of some religious sects (mostly Baptist and Methodist). The Second Great Awakening was responsible for many social reform movements such as the abolitionist crusade, women's rights movement, prison reforms, education reforms, temperance (anti-alcohol movement), asylum reforms and various rural utopian religious communal movements.

For the most part, Georgetown slaves did not rise up in violent conflict against their masters, but they did offer protests to being enslaved. Slaves malingered, or pretended to be sick, in order to get out of work. This action was more common among women than men. Some slaves performed their work poorly as a form of passive resistance and made sure that their task was not performed correctly. Other slaves protested by stealing from the master. Stealing was the most widespread form of protest among the slaves. Other less frequently practiced forms of protest were arson, tool breakage or sabotage of the master's property and the most serious of all: running away. Of course, running away meant avoiding not only the master but also the slave patrol with its dogs, horses and whips. Slaveholding states mandated that all adult men (whether they owned slaves or not) participate in slave patrols.[256]

In 1823, as a result of Denmark Vesey's failed slave insurrection in Charleston, the State of South Carolina passed many laws to regulate the way planters treated their slaves. Under these new laws, slaves could no longer buy their freedom, be taught to read and write, marry or have guns, among other regulations. However, tucked away in their country plantations, many planters broke these laws. One good way to control slaves, train them not to rebel and obey their masters was by teaching them that obeying God's laws was also to obey their master's laws on the estate.

The planter class in South Carolina's Lowcountry (as a group) honored, protected and respected women as part of the code of chivalry. They also appreciated and promoted feminine beauty by dressing their wives and daughters in fashionable attire, but they did not agree with the national women's rights movement of the era. Generally, planters were not concerned about equality between men and women but did take steps to ensure that their daughters grew up to be proper plantation mistresses. Planters hired personal tutors for their children and later sent them to finishing schools before sending them away to college.

Generally speaking, house servants went to church with their masters, who were usually Episcopalians, which made them practitioners of the same faith. Concerning field workers, many masters supported and encouraged the "missions to the slaves" provided by the protestant churches, including the Baptist, Methodist and Episcopal faiths. Slaves accepted Christianity, but rather than making a complete conversion, they blended Christianity with their own preexisting religious beliefs. The results of this blending of religions led blacks to focus on the salvation and redemption of Christianity, while whites focused on sin and damnation.[257]

In the Waccamaw region of Georgetown, this mission to the slaves was led by Reverend Alexander Glennie of All Saints Episcopal Church. Besides giving church sermons to the planters at their home church of All Saints Church and the sanctuary constructed near the Pawleys Island summer homes, Reverend Glennie preached to the slaves at thirteen chapels once a week at each of the following plantations: Woodbourne, Laurel Hill, Brook Green, The Oaks, Litchfield, Waverly, Midway, True Blue, Hagley (where the chapel featured stained-glass windows), Friendfield, Sandy Knoll, Cedar Grove and Mount Arena.

In 1844, Reverend Glennie published a collection of twenty-four sermons in his book *Sermons Preached on Plantations to Congregations of Negroes*. Within his sermons, Glennie preached of basic Christian stories such as the Garden of Eden, Passover, Good Friday and Easter. Reverend Glennie also taught Christian values such as humility, morality, purity, forgiveness and reliability. He entertained with stories of Jesus Christ's miracles, including curing leprosy, healing the sick, giving sight to the blind and raising Lazarus from the dead.

Reverend Glennie preached messages concerning slaves in relation to their masters. He aimed to appeal to and help the slaves by teaching that Christianity, in its basic form, instructed believers to fear hell, to model their lives after their savior who lived his entire life in poverty and to look forward to inheriting everlasting life as a reward for obeying God's laws. On at least one occasion, he said, "Masters are taught in the Bible, how they must rule their servants, and servants how they must obey their masters." On another occasion, Glennie taught slaves to

> *not attend to your work only while your master's eye is upon you; but remember that the eye of your heavenly Master is always upon you…be obedient to them that are your masters according to the flesh, with fear and trembling, in singleness of your heart, as unto Christ; not with an eye service as men pleasers; but as the servants of Christ, doing the will of God from the heart; with good will doing service, as to the Lord and not to men; knowing that whatsoever good things any man doeth, the same shall he receive of the lord, whether he be bond or free.*[258]

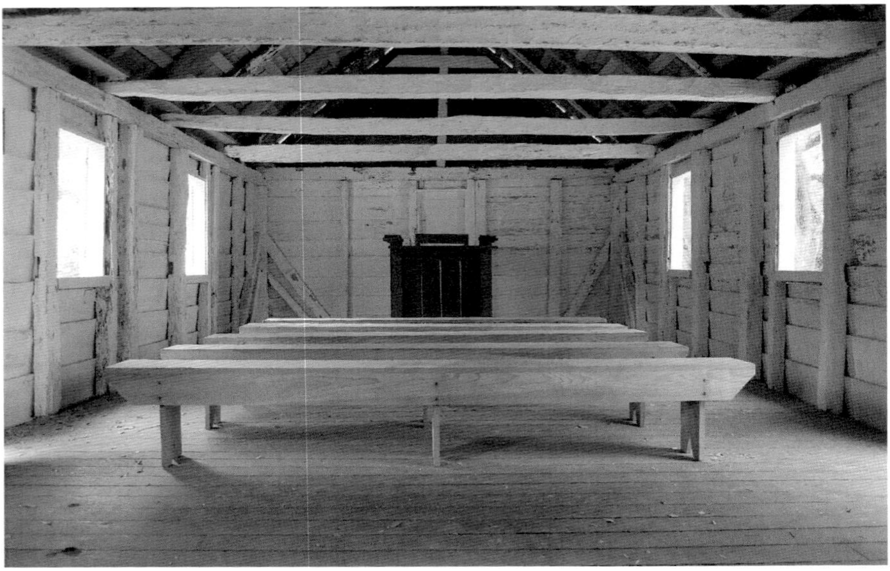

Mansfield's restored chapel contains simple benches much like those that were here when it was an operating church.

Other chapels existed on several plantations on the Black, Sampit and Pee Dee Rivers. Slaves on other plantations, such as those on Joel Roberts Poinsett's White House Plantation, held Christian services in a wagon shed or barn.[259]

Slave-led church services were filled with singing. Frederika Bremer, a visitor to South Carolina during the antebellum period, explained the service in a slave chapel: "They sang so that it was a pleasure to hear, with all their souls and with all their bodies in unison; for their bodies wagged, their heads nodded, their feet stamped, their knees shook, their elbows and their hands beat time to the tune."

Slave services were often so physically involving that slaves found back rails on their church pews to impede their services because it did not give them enough room to pray. Frederick Law Olmsted concluded that during services, slaves "work[ed] themselves up to a great pitch of excitement, in which they yell and cry aloud, and, finally shriek and leap up, clapping their hands and dancing, as is done at heathen festivals."[260]

Persuaded by religious zeal and concern for the salvation of their slaves' souls, and perhaps as a response to the northern abolitionists' claims that slaves were mistreated and kept in a state of ignorance, some planters and their wives brought great changes to their slave communities. Although made illegal in 1834, some planters, such as the Parkers and John D. Magill, taught their slaves to read and write. Fifteen freedman schools operated without government interference or funding between 1800 and 1860.[261]

As a general practice, Lowcountry planters during the antebellum period provided their slaves with watertight cabins with raised wooden floors and windows, with gardens behind the house and flowers in front. In most cases, the houses were better constructed than the homes of Russian serfs, Irish peasants and even most British and American industrial workers of the antebellum era.[262]

According to several visitors to the area during the antebellum years, slave villages were clean with well-constructed cabins in good repair. Slaves made their bedding out of wool, where available, while others simply used rice straw. Charles Mackay, editor of the *Illustrated London News*, toured the United States and recorded his experiences and findings in his book *Life and Liberty in America*. He wrote:

> *In the village there were a hospital, an infirmary for the sick, a chapel, where every Sunday Divine service was performed by a missionary, allowed to have access to the slaves upon condition of not preaching freedom to them, and a nursery, where the young children from the earliest age upward to fourteen, were taken care of during the absence of their parents in the fields.*[263]

Fellow planter and physician Dr. James Ritchie Sparkman wrote a letter to another planter explaining his slave village, which bore a striking resemblance to Dr. Parker's village at Mansfield. Dr. Sparkman wrote:

> *Mine are all well framed buildings 18 by 22 feet, of best material hewn or sawed frames, milled weatherboarding, covered with best Cypress shingles, raised 2 feet from the ground, flooring closely joined, glassed lights* [probably windows] *to each room, and large fireplaces or chimneys made of a composition of clay, sand and tar, as a substitute for brick…Each house contains a hall and 2 sleeping apartments and is intended to accommodate an average of five people, or one family.*

The Parkers, Westons, Sparkmans and many others provided their slaves with well-built homes that were raised off the ground and had wooden floors instead of dirt floors. Others were raised on brick piers

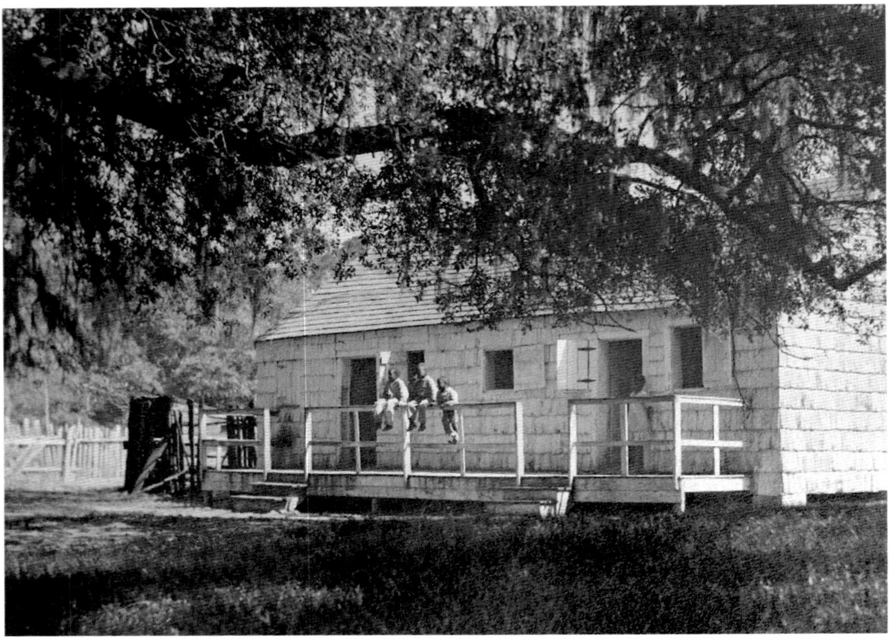

Anonymous children on the front porch of a cabin, circa late 1930s.

two feet or more high. One visitor claimed that all slave huts were built this way by the late antebellum period.[264]

Although it was technically illegal, planters often encouraged marriage as a way to secure a labor force for the next generation. In some instances, masters sent out invitations to the weddings and provided the married couple and attendees with cake, turkey, ham, molasses, coffee and other foods. Sometimes, the plantation mistress would give a slave woman one of her dresses to wear or even a new dress of her own. As in the case of favorite house servants, often a white minister performed the service. In some cases, slaves were even given a three-day holiday as a honeymoon, housekeeping articles and, sometimes, their own cabin for starting a family.[265]

Slave marriages almost always included the ceremonious "jumping of the broom," signifying that the couple lived together. If either of the two individuals stubbed their toe on the broom, it meant that the couple would experience marital troubles.[266]

If a male slave married a woman from a neighboring plantation, the wife was called a "broad wife," and the children lived with her and became the property of the female's master.[267]

Some planters allowed slaves to receive other sacraments, such as baptism and confirmation, and even allowed some slaves to administer church services and lead prayer in slave village churches. Henry Wallace was a slave who performed Christian rites at slave funerals.[268]

During his visit to Horry and Georgetown Counties during the war years, Anglican reverend William Wyndham Malet worked with slaves in reciting their catechism and even witnessed circumcisions, the mark of Jewish tradition. He also witnessed a slave named Jemmy leading others in prayer during the war years. Jemmy prayed:

> *O Lord, in whose palm of his hand be the waters of the ocean—who can remove the mountains—who weighs the earth in a balance—who can still the waves of the storm—who can break the pines of the forest—who givest us a land of rivers of waters—O Jesus! Who died on the cross of us—O forgive us our sins; O help us in this time of trial and need. Protect our massa far away; protect our massa far away; protect our brothers "Hector" and "Caesar" with him; defend us now we are away from home; defend our friends and relatives at home.*[269]

Even as slaves rowed boats down the rivers, they could be heard singing fine solemn hymns. Women's voices mingled, improvised by boatmen. The hallelujah was prolonged to give the singer time to remember or improvise the next line:

*The Jews killed my Jesus—Hallelujah*
*Upon the cross they stretched him—Hallelujah*
*They laid him in the Sepulcher—Hallelujah*
*Then early in the morning—Hallelujah*
*Came Mary and Joanna—Hallelujah*
*And asked for Master Jesus—Hallelujah*
*Two angels were a-sitting—Hallelujah*
*Where he had been lying—Hallelujah*
*Jesus was a-standing—Hallelujah*
*Hard by in the garden—Hallelujah*
*Mary did not know him—Hallelujah*
*And said, "Where has thou laid him?"—Hallelujah*
*"Mary, don't you know me?"—Hallelujah*
*Then said she, "Rabboni"—Hallelujah*[270]

Mr. John H. Tucker, proprietor of Litchfield and Willbrook Plantations, was clearly concerned about the salvation of his slaves' souls and linked their church attendance to their weekly allotments:

> *Mr. Tucker informed me that he required the overseer to call roll after service, and if any Negro failed to answer his name, who was not on the nurse's sick list for that day, he was cut off from the weekly allowance of bacon, sugar, molasses, or tobacco. The forfeiture was found sufficient to ensure a punctual attendance. The hint may be taken by our planters, who sometimes find reluctance to attend religious services where they have established them on the plantations. In such cases, Mr. T's plan is a wise one. It attacks the African in his weakest point, and appeals to this appetite for the good of his soul.*[271]

In this close-knit, patriarchal society, men controlled everything from economics to politics and social order. Perhaps the proprietors of this civilization encouraged education and Christianity for their slaves as a means of self-gratification and personal triumphs in qualifying slavery as a positive good for Africans, as one planter clearly pointed out in his address before the citizens of All Saints Parish at Wachesaw Plantation on July 4, 1857:

> *For more than 200 years portions of this people* [Africans] *have been under our subjection, and we point with conscientious satisfaction to the effect already pronounced. The wily, blood-thirsty savage has become the obedient, civil, and Christianized servant; like the demoniac once cured on the Lake of Tiberias, he is now clothed and in his right mind. A moral revolution which once appeared impossible, or only to be wrought in an incalculable number of ages, has by the gracious providence of God, been brought about by removing this wild race to another country and by subjecting it to a restraint which at once enriches a great nation and elevates it.*[272]

Whether Christianity was used as a form of control to create greater subordination among slaves and teach them respect and obedience as a means to explain slavery as a positive good or as a means to save the souls of the slaves, one thing is for certain: Christianity was brought to slaves. Reverend Malet noted the successful impact that the Christian mission and education had on the slaves of one plantation:

> *There are nine women and men superannuated* [retired because of age or infirmity], *all comfortably housed and cared for: several of the boys and men can read and write; the girls when young can get over the rudiments of reading, but have a most extraordinary inability to proceed; yet by viva-voce* [spoken] *teaching they get up their catechism* [basic principles of Christianity in question-and-answer form] *very tolerably, and also scripture history; and many answered my questions better than our poor children do in most places.*[273]

Like most traditional African American cemeteries, the cemetery at Mansfield is a very simple resting place for the deceased. The slave funeral represented life's climax and completion. Friends and loved ones of the deceased gathered, prayed and sang for hours before they processed to the cemetery. The funeral procession usually took place at night. Loved ones sang melancholy songs on their walk to the grave site. Sometimes, slaves marched around the grave site and beat a drum and shouted. After the ceremony, slaves sang joyous songs. To many slaves, death marked a reunion with deceased kinsmen and women, as well as a release from earthly ailments.[274]

Unlike the typical European American cemetery, which is carefully laid out in family-based plots, constructed on prominent hill tops, well tended and filled with commercial stone markers, the African American cemetery lacks clear organization, is kin based, is constructed on low swamp lands or hidden in the woods, has a variety of markers and features living memorials (such as flowers and bulbs) or other personal objects of the deceased (such as bowls, glasses, plates or seashells) instead of commercial headstones.[275]

During the colonial period, most slaves constructed their own bowls, drinking vessels and pitchers and cooking pots on the plantations. Most of this low-fired pottery is known as colonoware. This African tradition of making pottery is identified as brown or black undecorated items, depending on oxidation during the firing stage. At Mansfield, Professor Jim Michie's archaeology survey uncovered many fragments of colonoware. Mr. Michie's crew recovered most of the shards in the colonial and early antebellum period slave village (under the ownership of Susannah and Mary Man, John Man Taylor and Reverend Maurice Harvey Lance) but also found pieces in the location of the late antebellum period slave village (the Dr. Francis S. Parker village) and near every structure on the estate.[276]

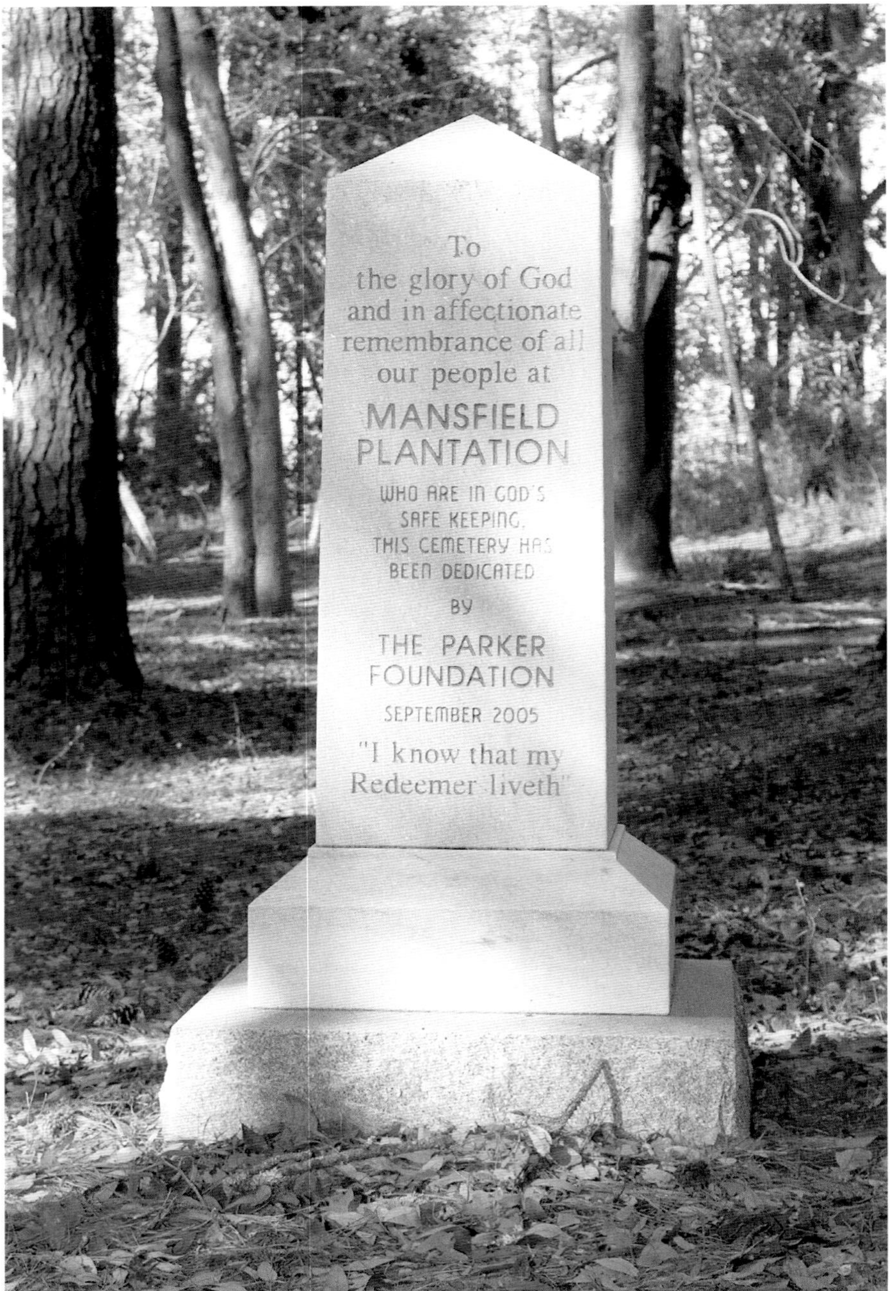


The cemetery marker is a welcoming obelisk structure designed to pay respects to all those who labored at Mansfield. The cemetery is open to descendants of Mansfield and maintained by the author of this book and his students.

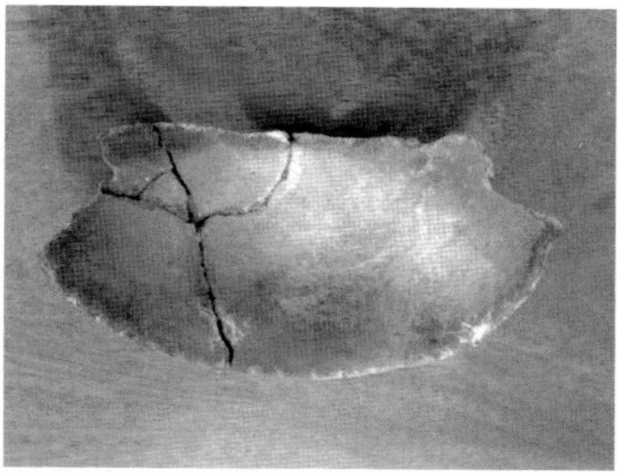

The "X" mark on this colonoware piece suggests that the craftsman or woman who made the bowl was in mourning when he or she formed the vessel.

One interesting piece of colonoware recovered by Susan McMillan and the Waccamaw Diggers during the second archaeological excavation of Mansfield Plantation was the partial remains of a colonoware vessel incised with an "X" marking. This West African religious marking is common, as is the Congo cosmogram, which represents the sun rising over the earth to return to the underworld at night. The people of the Congo believed in the existence of two worlds (the living and the dead) separated by the river Kalunga. To the people of the Congo, death is simply a transition of life from one world to the next. African American potters preserved and continued these traditions and religious beliefs in South Carolina.[277]

At Mansfield, Professor Michie found another crude ceramic in the pre-Parker-era slave village. River-burnished pottery was made from local clay tempered with small particles of sand and was occasionally painted with black and red lines. Mr. Michie reported finding a few shards of this pottery decorated with red lines and dots, but they were too small to determine their form or function.[278]

During the antebellum period, the use of colonoware and river-burnished pottery disappeared as planters such as John Man Taylor, Reverend Maurice Harvey Lance and Dr. Francis S. Parker purchased pearl wares and white wares for distribution to their slaves. Increasingly, more and more planters provided inexpensive store-bought ceramics. If a piece of their own fine porcelain broke, they often distributed the rest of the set to their house servants when they purchased a new set. These valuable objects were symbols of high status among slaves. Domestics had the option of keeping the dinnerware or trading it to field workers in the community for items they wanted, creating an informal slave economy. The informal slave economy led to economic success for some slaves and further contributed to class differences among them.[279] Today, bits of commercially produced ceramics litter the landscape of Mansfield Plantation.[280]

In the old slave village, archaeological excavations uncovered a few pieces of overglazed enameled Chinese pottery, three types of earthenware, lead-glazed slipware (produced between 1670 and 1795), creamware (1762–1820) and undecorated whiteware (1820–1900). The excavations recovered more than twice as much colonoware as commercially produced tableware. The mean ceramic date of all manufactured items in the old village is 1768, compared to a mean ceramic date of 1786 at the presumed kitchen area of the old settlement pattern. At the old kitchen area, earthenware such as undecorated whiteware and lead-glazed slipware dominate the cache, while colonoware represents about a third of the total findings. These findings agree with the hypothesis that planters handed down ceramics to house servants.[281]

The area around the kitchen constructed under the Dr. Francis S. Parker ownership reveals a very different pattern. While undecorated earthenware still dominated the cache, a large amount of white, undecorated ironstone stoneware was recovered. Most importantly, only a few shards of colonoware were recovered from the area, and the mean ceramic date of the site is 1844.[282] These findings support the thesis that the house servants received the planter's discards.

# Chapter 7
# Planter Lifestyle

*The beautiful and fertile Lowcountry of our state is the seat of annual and endemic visitations of disease, which we are accustomed to attribute to Malaria…there are few plantations which admit of permanent residence; the whole region being pervaded by a pestilential infection, almost unfailing in the excitement of Fevers, Intermittent and Remittent, during the Summer and Autumn.*
*—S.H. Dickerson*[283]

By 1840, the rice planters in the Georgetown District were gentlemen of leisure. The creation of the rice plantation system was nearly complete, as very few plantations were constructed during the late antebellum period. Planters hired overseers and, to a lesser extent, drivers to manage fields and laborers. Planters employed factors to market and sell their crops and bankers to handle their money. The job of planters themselves was to provide for their families and slaves in great paternalistic fashion. With their plantation concerns handled for them, many members of the final generation of antebellum rice planters lived their lives traveling, governing politics and leading lives of leisure mandated by the seasons.

Each year at the end of May, out of fear of contracting malaria, commonly referred to as "country fever," the planting families moved away from their plantations and did not return until the first week in November.[284] They spent time with relatives or friends in other parts of South Carolina, North Carolina or other states. Some spent time abroad in Europe. By this time, the resort at North Island had been devastated by the hurricane of 1822, and planters built new retreats. The most popular in-state retreats of the Georgetown planters were Pawleys Island, Murrells Inlet, Charleston and a small community on the Pee Dee River called Plantersville.

The most celebrated of these summer retreats were Pawleys Island and Murrells Inlet. Located at the seashore and marsh front respectively, these twin communities continue to be among America's oldest resorts. For the Pee Dee families, who had to cross the Waccamaw River in order to get to the beach, getting to their summer refuges was a long, hard trek. But for the Waccamaw families, whose land stretched from river to sea, it was closer and more convenient. One woman recalled her family's annual pilgrimage to the coast: "All of our belongings, servants, horses, cows, furniture, were loaded on to lighters and propelled seven miles through broad rivers and winding creeks to Waverly Mills, where they were disembarked and traveled four miles by land…but when we reached this paradise on the Atlantic Ocean we felt repaid for it."[285]

As author Lawrence Fay Brewster stated, "Waccamaw River planters had the advantage of having retreats within easy reach of their plantation dwellings, either on the beach heads of their own lands, which stretched from river to sea, or at nearby resorts, to which they migrated in summer with their families, friends, pastors, teachers, and servants."[286]

Planting families traveled with everything that they would need for the six-month period away from home. Families spent weeks packing. Resembling small armies as they traveled, planters annually moved everything from kitchen utensils and clothing to pianos to their summer refuges. Mrs. John Weston explained why she took so many servants with them to the beach and to Charleston: "We can not possibly separate husband and wife for six months; so Harry, the coachman, has to have his wife and children, and the same with the cook, and the butler, and the laundress, until we are actually moving an army every time we move."[287] Mrs. Paul Weston concurred, "We have to take fifty individuals with us in the move, I mean children and all."[288]

Pawleys Island, a resort community since the late 1700s, began to expand during the 1820s due to the growing wealth and leisure time of the planters. When the hurricane of September 1822 destroyed most of the older buildings, the residents quickly built new homes on their properties. To accomplish the rebuilding, planters sent skilled carpenter slaves with cypress timber by boat to restore the community. One head carpenter slave cut the materials for the family's summer home and labeled each piece at their home plantation so it could be quickly assembled at the beachfront property.[289] Several planters built notable homes at the summer resorts, including Plowden C.J. Weston, Dr. E.T. Heriot (Woodland), Colonel Joshua John Ward (Magnolia, Retreat), Robert F.W. Allston (Canaan) and Jacob Motte Alston (Sunnyside). Dr. Andrew Hasell, a full-time resident of Pawleys Island, resided at Cedar Grove. Most of these houses had wide porches facing the southwest and southeast, the directions of the prevailing winds, while some had piazzas that encircled the entire structure.

By the 1840s, Pawleys Island, a strip of land a quarter of a mile across (at its widest point) and less than four miles long, was clearly more popular than Murrells Inlet.[290] A northern visitor to Pawleys Island in 1851 described the resort very descriptively. She wrote, "I rode over to one Pawleys Island which is three miles long, with a fine beach on the ocean side, some 6 or 8 houses, plenty of trees &c separated by a creek from the main land which we forded."[291]

For summer recreation, the planters created a social club, the Hot and Hot Fish Club, where they dined on extravagant meals and entertained themselves by fishing, hunting, racing their horses and playing ten pin and billiards. The hunters had small clubhouses made of pine logs constructed at central locations in the forest where they met to take breaks, sip on fine wines and enjoy their kill and catch, which was prepared by their slaves.[292]

While the planters entertained themselves with their convivial gatherings, their families kept amused around the summer home. Adele Petigru Allston wrote to her husband addressing her summer routine while he was away on business: "I walk the causeway every morning after breakfast, and stroll upon the beach and sand hills in the afternoon."[293] During their summer vacation, the children were allowed to enjoy the early morning hours; however, they were required to keep up with their studies. After dinner, they enjoyed "swimming in the surf, horseback riding on the beach and, on rainy days, the great privilege of reading fiction."[294]

Whether at home or at the beach, Georgetown planters involved themselves in religious activities. The Episcopal faith, the religion of the planter class, boasted three fine churches in Georgetown. In town, the elite worshipped at Prince George Winyah Church. At home in the country, they attended either All Saints Church on the Waccamaw or Prince Frederick on the Pee Dee River. While at the beach, plantation owners and their families flocked to All Saints Parish Episcopal Church at Pawleys Island.

In July 1832, the planters and their wives, in conjunction with Reverend Alexander Glennie, the pastor of All Saints Church, formed the All Saints Sunday School Society. The school's function was "to afford religious instruction to the children of the Parish, according to the Protestant Episcopalian Church Sunday School Union."[295] The society created a constitution with six rules and held elections each year for the offices of president, vice-president, secretary and treasurer, as well as three directors.[296]

In 1838, the All Saints Summer Academy, directed by David D. Rosa, opened.[297] By 1843, the school served both male and female students and offered "a complete English education" to anyone who could afford tuition, which was $20.00 per term for students over twelve years of age and $10.00 for students under twelve. Latin, Greek, French, music and drawing classes required additional fees. Classes began on January 2 of each year, and the school expected the students to provide their own furniture. Board and "washing" fees cost an extra $13.50 per month.[298] In 1854, Peter Waities Fraser donated three hundred feet of land at Pawleys Island for the purpose of building a summer rectory. The erection of this home for Reverend Alexander Glennie was the church's final expansion into the beach communities during the antebellum years.[299]

Many Pee Dee and Black River rice growers, including John Harleston Read II and Maurice Harvey Lance, and especially the individuals who could not afford good overseers, simply moved farther up the Pee Dee River to the safety of the pine lands during the summer months. From there, planters could check on their estates and still enjoy good health during "the sickly season." In 1852, these summer residents incorporated this 350-acre resort and formally named it Plantersville.[300] Even after the area began to attract outsiders, the resort remained the preferred vacation spot of the Black River and Pee Dee families due to its proximity to their homes.

The planters returned home during the first week in November. Winter was the time when planters entertained one another by holding large lawn and garden parties during the day, as well as dinnertime and evening parties. Although the men had the Planters' Club on the Pee Dee and the Winyah and All Saints Agricultural Society for amusement during the winter months, they primarily regarded this season as the time for family gatherings, weddings and holiday celebrations.

At one such wedding party, in the winter of 1849, Joshua John Ward gave a dinner party at Brook Green in honor of his daughter Penelope's marriage to Dr. Arthur Belin Flagg. The Heriots, Belins, Petigrus, LaBruces, Vaux, several of the Allston and Alston families, Izards, Pringles, Poinsetts, Reads and some of the Westons and Fords were in attendance. According to one guest, "The party was kept up till the we sma' hours, and there was one who greatly enjoyed a regular old fashioned country dance to the music of sundry country fiddles." Five years later, Georgeana Ward married, and the family held a similar wedding ceremony.[301]

When recalling a plantation wedding, one Santee River resident wrote:

> *I let myself loose to the enjoyment of the hour; danced all night, spent the next day in walking, riding and other sports to which the party gave themselves up before dinner and reappeared in the evening ready to dance with the most determined. At that time the wedding festivities always continued two nights, nor were the guests expected to leave the house until the third day.*[302]

Another gentleman from the Georgetown countryside recalled the formality of plantation weddings in his memoirs. He wrote, "Weddings were celebrated on these river plantations and the guests came from as much as twenty miles in boats rowed by stalwart negroes. They remained for the night and continued the festivities the next day. On the return voyage the negro oarsmen kept time to their oars with improvised songs in honor of the bride and groom."[303]

During the holiday season, the planters displayed their true patriarchal spirit. Although the roads were unpaved, families from throughout the Georgetown countryside, as well as from the city, traveled about visiting friends and kin. Planting families decorated their manor homes with holly. Treats such as eggnog, ginger bread and assorted cakes were often served in abundance.

Christmas in the Georgetown District was a time of celebration for all Christians, whether master or slave. At most plantations, the celebration lasted for three days, and the only work done on the plantation during that time period revolved around necessities such as preparing food for the family and tending to the livestock. At Chicora Wood Plantation, home of Robert F.W. Allston, Allston himself set off fireworks on Christmas Eve for everyone's delight, and his daughters helped all the children hang their stockings because even slave children expected a visit from St. Nicholas.

For a week before Christmas, the whole plantation unit was busy preparing gifts for one another. On Christmas morning, the family and their slaves feasted and exchanged tobacco, fruit, candy, dolls and handkerchiefs. Then the whole plantation community gathered in front of the manor house and on the front piazza of the house. Slaves danced to music that they played on fiddles, triangles, jaw-harps and assorted sticks and bones. The dancing and partying continued throughout the three-day holiday, with dancers and musicians taking turns on the piazza while others feasted on beef and pork and drank alcohol.[304]

On Christmas Day 1856, the year that Allston won the gubernatorial race in South Carolina, there was a special celebration at his household. Mrs. Allston recalled, "The next morning there was a great demonstration, the negroes coming up as usual to greet their master the Gov. They made a great noise and drank the Governor's health in many a stout glass of whiskey."[305]

On some plantations, the planting family and their slaves went caroling. Many planters handed out list tickets to slaves so that they could visit their friends and family members at nearby plantations. Christmas was a popular time for slave marriages. The newlywed couples were usually toasted by their owners and granted an additional three-day vacation from their duties.[306]

New Year's Eve was another holiday that earned the whole community a three-day period of rest, visitation and celebration. On New Year's Day, some masters gave annual rations of clothing, blankets and shoes to the slave families.[307] Meanwhile, the planting families began to turn their attention toward the "social season."

Set in Charleston, the social season was a chance for the planters throughout the state to come together for communal discourse. The season lasted from the middle of January until Palm Sunday (the week before Easter). Some who owned houses in Charleston were Dr. Francis S. Parker, John Harleston Read II, Joshua John Ward, Francis Marion Weston, William Algernon Alston, John H. Tucker, Robert F.W. Allston, Charles Alston, Jacob Motte Alston and nearly twenty others. Reverend Maurice Harvey Lance, perhaps because he performed so much itinerate preaching in the upstate, preferred to rent a house in Charleston.[308]

The gentlemen took their ideas of liberty and secession with them on their vacations. George C. Rogers Jr. contends in his book *The History of Georgetown County* that the presence of these planters from throughout the Lowcountry influenced the Charleston merchant class to vote for secession in 1860. Rogers claims that Lowcountry planters, primarily those from Georgetown and Beaufort, aided by Robert Barnwell Rhett, the fire-eater politician, journalist and proprietor of the *Charleston Mercury*, dominated the city and persuaded Charlestonians to seek independence from the United States.[309]

Of course, some planters chose to travel throughout the United States or abroad during the winter months, but Charleston was the place the planters visited most frequently. They relaxed with leisurely walks in public pleasure grounds such as the Battery Promenade, White Point Gardens and the Rutledge Avenue Pond. While at the Battery, those with an appetite for sweets could enjoy refreshments such as ice cream and pastries. In the evenings, the sounds of brass and woodwind instruments or vocalists, pianists and violinists entertained the crowds strolling or riding in the cool night air viewing the harbor.[310]

The Georgetown elite also intermingled with Charleston society by attending weddings, musicals, soirees, debutante balls, dances, masquerades and parties sponsored by such prominent groups as the St. Cecilia Society, the Cotillion Club and the Jockey Club.[311] The organizations held their gatherings at exclusive auditoriums such as the South Carolina Institute Hall on Broad Street, Military Hall on Wentworth Street, Hibernian Hall and the New Charleston Theater on Meeting Street.[312]

Of all the societies in Charleston during the antebellum period, the St. Cecilia Society held the most formal balls. Planting families received invitations to three or four balls each week. The St. Cecilia Society held a ball every ten days. Many women wore French designer dresses to the St. Cecilia Society balls, which were the "most exclusive and elegant balls of them all."[313] The balls centered on social intercourse, fine wines, food and dancing. Popular dances of the era were the waltz, the polka and the maruka, collectively referred to as "round dances." Proper ladies were not expected, and some were not permitted, to partake in "fast" dances.[314]

In 1762, the women of Charleston founded the St. Cecilia Society as a musical organization; however, by the 1840s, the society had become more of a cotillion club than an organization to provide quality music to the socially refined.[315] The society held its largest ball annually during February (just before lent) in Hibernian Hall on the west side of Meeting Street. A contemporary remarked, "The membership remains exclusive and its affairs somewhat secret."[316] Even today, the club remains secretive and its historical records and membership lists remain off-limits to non-members. Throughout the year, the merchant class reigned in Charleston, but during the second half of January, February and into March, the city was in the hands of the planting class. In this unofficial capital of the Lowcountry, the planting families displayed their wealth in flamboyant fashion.

The frock coat dominated the wardrobe of the 1840s gentleman. Meant to be worn with knickers and full shirts, it featured a deep collar and wide lapels and was often edged with braid. Another popular style, more formal in design, was the patterned or embroidered waistcoat worn with tight-fitting trousers that fastened under the instep. Men's accessories of the period were the cravat necktie and gutta percha cane. During this decade, ladies wore hooped dresses with sloping shoulders and tight sleeves. Women's accessories included ermine muffs, tasseled handbags and parasols for outdoor events. Their tightly fitting, flower-trimmed bonnets had deep crowns and large brims that framed and complemented their faces. During the evening, ladies wore dresses with provocatively low-cut bodices and feathers or pearls in their hair.[317]

During the 1850s, although fashions did not change much, the number of garments required for gentlemanly attire proliferated. Besides the morning frock, dress and overcoats, men wore paletot cloaks with contrasting collars and slit armholes. The popularity of cravats declined, and proper neckwear became a bow or shoestring necktie. By 1856, the introduction of the caged crinoline had relieved the weight of women's petticoats. Layered flounced skirts became popular. Ladies still preferred dresses with sloped shoulders and hats that tied firmly under their chins; however, bonnets no longer had deep crowns and brims. Women adorned their "round" hats with long, wide ribbons and wore their creations farther back on their heads. Very small and highly decorated parasols were the rage. Handbags were also minuscule, and neither served much purpose. One prevailing purse design was the "miser," which had fringe at one end and tassels at the other, while another was made of netting and beads. During the evenings, ladies wore snake-like jewelry and headwear made of pearls and flowers set far back on their heads.[318]

The Charleston Jockey Club, founded in 1793, was part of a fifteen-track circuit of jockey clubs that existed in the state during the antebellum period. The Charleston Jockey Club held its races at the Washington Course near Charleston.[319] The Charleston races occurred, by rule, on the Saturday, Sunday and Monday preceding the first Wednesday in February.[320] One

Georgetonian who frequently visited the track noted, "The Ladies of Charleston always turned out in full feathers on these occasions, and the grandstand was filled with lovely ladies who took intense interest in the horses which were then owned by Gentleman of note."[321]

At the Washington Course, the races began each day at 1:00 p.m. The first day's races were four-mile heats; the second day, three-mile heats; and the third day, two-mile heats. Daily entry fees for members were twenty dollars, fifteen dollars and ten dollars. Non-members paid double. The daily purses varied depending on the number of entries in each race.[322]

The Jockey Club required a forty-dollar annual fee from its members, which defrayed the cost of the races, to maintain the track and to pay for dinner parties such as the one held annually on the Wednesday following the races.[323] The club held its annual ball on the Friday following the races. This event signaled the end of the state's professional race season.[324] According to one Georgetown race fan, the ball was "the largest and grandest" of them all.[325]

Usually, at the end of race season, the planters returned home to their manor houses in the countryside to enjoy their estates and the closeness of their small yet extremely affluent social organizations until May, when the cycle started again.

## Chapter 8

# Dr. Francis S. Parker, Politician and Provost Marshal

*The safety of the people must be the supreme law.*
—*Roman proverb*[326]

On November 8, 1860, the *Charleston Mercury* wrote, "The tea has been thrown overboard—the revolution of 1860 has been initiated." The newspaper called for "minute men" organizations throughout the state to prepare for war.[327] The United States citizens elected Republican Abraham Lincoln as president. Immediately following the election, on November 13, 1860, the South Carolina General Assembly in joint session ratified an act calling for a "convention of the people to convene in Columbia on December 17." The election of delegates from each voting district in the state took place on December 6. All of the Unionist candidates and strong opponents of secession were overwhelmingly defeated. One such detractor from secession was Charleston attorney James Louis Petigru (brother-in-law of Robert F.W. Allston). Petigru warned the people of South Carolina when voting on their candidates for the convention that "South Carolina is too small for a Republic and too large for an insane asylum."[328] Nonetheless, the time for compromise had passed, and South Carolinians were ready for action.

The Georgetown planters, like many other planters in South Carolina, felt that the rights of their state had been violated far too often and voiced their discontent with the Union by declaring a Second American Revolution. Prince George Winyah Parish elected Samuel

Taylor Atkinson, Benjamin Faneuil Dunkin, Dr. Alexius Mador Forster and Dr. Francis S. Parker.[329] These were not the wealthiest men in the parish, nor were they the most politically accomplished; rather, they were considered moderates. Former governor Robert F.W. Allston owned several plantations and over 600 slaves,[330] but he was not elected to the convention because he was an extreme secessionist (commonly referred to as fire-eaters). John Harleston Read II, Parker's brother-in-law, was a career politician and owner of 511 slaves and several plantations, but he was not elected because he was still considered a Unionist at that time.

Samuel Taylor Atkinson was a forty-two-year-old who owned 18 slaves. He was a lawyer, not a planter. Atkinson owned only $4,000 worth of real estate and $23,000 worth of personal property. Chancellor Benjamin Faneuil Dunkin, originally from Massachusetts, was a judge in Georgetown. He was the proprietor of Midway on the Waccamaw, where 135 of his slaves lived; Bienvenue Plantation (French for "welcome") on the Pee Dee River, where he housed 97 laborers; and a home in Charleston at 47 Warren Street, where 4 house servants resided.

The people also elected Dr. Alexius Mador Forster. Forster, forty-four years old, was the holder of $50,000 worth of real estate and $202,000 in personal property, including 273 laborers who toiled at Friendfield Plantation on the Sampit River. Friendfield was the enormous estate of Francis Withers, a conglomerate plantation that included Friendfield, Mount Pleasant, Midway and Canaan Plantations. Forster came into the estate by his marriage to Elizabeth Hunt Warham, the stepdaughter of Robert Francis Withers. All of the slaves and the land actually belonged to Elizabeth's mother.[331]

All Saints Parish was divided by voting district lines, thus the Horry District and the Georgetown District each sent two representatives. Of the two representatives from the Horry District, thirty-five-year-old Thomas W. Beaty was an attorney and a merchant who lived in the village of Conwayborough. Beaty owned 11 slaves. William J. Ellis was a fifty-six-year-old farmer who owned $3,000 in real estate and $6,925 in personal property, including his 2 slaves. The two representatives from All Saints Georgetown were Benjamin Esson Sessions, a forty-five-year-old farmer who owned 6 slaves, and John Izard Middleton, a sixty-year-old planter who owned $15,000 worth of real estate, including Rice Hope and Blue Hall, known collectively as Bannockburn Plantation, as well as Clermont Plantation on the Waccamaw River and $40,000 worth of personal property that included 318 slaves.[332]

Collectively, 169 men attended the convention, which was held on December 17, 1860. Nearly all of the men owned slaves. Almost half of the men owned at least fifty, and twenty-seven had over one hundred or more slaves. Most were planters, but others were slave-less farmers, manufacturers, attorneys, judges, physicians, professors, clergymen, brokers, solicitors, a railroad president, a commercial merchant, president of an insurance company and even a schoolteacher who owned neither slaves nor property in South Carolina.[333] Interestingly enough, the group who gathered to decide the question of secession was probably more diverse in wealth than the Second Continental Congress that gathered in Philadelphia in 1776 to declare independence from England.

On the first day of the convention, John A. Inglis of the Chesterfield District introduced the resolution that South Carolina should secede from the United States at once. Before a committee to draft the document could be named, the convention adjourned in fear of a smallpox epidemic. The delegation decided to reconvene at Institute Hall in Charleston the next day. The convention's president, David F. Jamison, named Robert Barnwell Rhett (Charleston District), James Chestnut Jr. (Kershaw District), James L. Orr (Anderson District), Maxcy Gregg (Richland District), Benjamin Faneuil Dunkin (Prince George Winyah Parish, Georgetown District) and W. Ferguson Hutson (Prince Williams Parish, Beaufort District) to a committee to draft an ordinance and appointed John A. Inglis as chairman of the committee.[334]

The convention reconvened on December 20, 1860, at St. Andrew's Hall in Charleston. There, in the early afternoon, Chairman Inglis rose to deliver the committee's report. He simply read a treatise written by W. Ferguson Hutson, the primary author of the Ordinance of Secession:

> *The People of the state of South Carolina, in Convention assembled, declare that the Ordinance adopted by us in Convention, on the twenty-third day of May, in the year of our Lord one thousand seven hundred and eighty-eight, whereby the Constitution of the United States of America was ratified, and also all Acts and parts of Acts of the General Assembly of this state, ratifying Amendments of the said Constitution, are hereby repealed; and that the union now subsisting between South Carolina and other states, under the name of "The United States of America," is hereby dissolved.*

Plain, simple and to the point—there was no need to debate. At 1:07 p.m., a roll call vote was taken by last name. The meeting ended at 1:15

p.m. South Carolina seceded by unanimous vote. Shortly thereafter, the convention agreed to prepare an official copy of the document on parchment paper to be signed by the entire delegation. The signing took place that evening (December 20, 1860) at Institute Hall. At 6:45 p.m., the delegates marched, in procession, from St Andrew's Hall to meet the state legislatures at the foot of Institute Hall's stairs. The three-thousand-person facility was filled to over capacity. The convention, followed by the South Carolina Senate and the South Carolina House of Representatives, entered the building. Governor Francis Wilkinson Pickens, the president and clerk of the convention, the president of the senate (William Dennison Porter) and his clerk, the Speaker of the House of Representatives (James Simons II) and his clerk and the Lutheran pastor (Dr. John Bachman) waited on stage for the delegations to enter the building. After a prayer from the pastor, convention president Jamison read the lrdinance. After a thunderous applause shook the building, the delegates took two hours to sign the document (called alphabetically) by election district.[335]

Outside Institute Hall, bells rang, crowds cheered, bands played, bonfires burned, cannons boomed on forts and rockets exploded past midnight. Parades, flag-waving and bell-ringing continued the next day.[336] South Carolina had finally declared independence; it was a long time coming.

When the delegates to the secession convention signed the Ordinance of Secession, they went home as heroes, the architects of the Republic of South Carolina. At the end of March 1861, the delegates came together again and voted 99–63 to reproduce two hundred copies of the Ordinance of Secession for $200. Evans and Cogswell (Charleston printers) had anticipated the order and had already printed them.[337] At the conclusion of business, each member of the delegation took his prize home with him. By then, South Carolina no longer stood alone; it was a member of a new nation: the Confederate States of America, born in discontent and soon to be baptized in war. The war, which pitched central authority against local autonomy, federal authority versus states' rights, industry against agriculture and North versus South, was an independence movement, not a civil war.

During the war effort, Dr. Parker continued to serve as a vestryman at Prince George Winyah Church and perform other civic responsibilities as he did before the war. However, the physician-planter turned politician also took on the tremendous responsibilities of provost marshal of

Georgetown. This military position gave him the authority to regulate the rights and privileges of all people in the Georgetown District—both military and civilian, free and chattel African Americans—and to punish lawbreakers with military authority. The provost was also in charge of upholding the morals of the military by eliminating drinking, prostitution and gambling near the encamped soldiers and near the vicinity of troops. He was expected to provide guards and escorts as needed for headquarters, military hospitals, jails, public and private property, civilian visitors in nearby encampments, incarcerated soldiers and civilians, prisoners of war and prisoners of special nature such as spies.

The provost was liable for combating enemy spying, espionage and sabotage. He was responsible for conducting and apprehending traitors and preventing the sabotage of trains, railroads and bridges by the enemy forces, and he was liable for the gathering of intelligence, as well as managing Confederate spies. The provost was also in charge of conducting espionage and misinformation against the enemy and retaining control over civilians and soldiers in the town.

Additional responsibilities included being in charge of supervising the movement of all military and civilian personnel by issuing travel passes and passports, as well as monitoring all shipping and transportation. The provost was in control of overseeing the impressments of local supplies for use by the military; overseeing the local slave labor, both impressed or volunteered; and keeping control over enemy prisoners, enemy deserters and recaptured runaway slaves and administering justice to the runaways.

The position of provost marshal meant maintaining the fighting forces, retaining command discipline, supporting intelligence/counterintelligence organizations and other miscellaneous duties. In maintaining the fighting force, the provost oversaw the recruitment of volunteers and enforced conscription among those who were less than enthusiastic about fighting for Southern independence from the Union. He controlled straggling troops who moved through the area, prevented desertion and ensured that furloughed men returned to their units after their furlough expired. He also hunted down and apprehended those who had deserted.

As provost marshal, Dr. Parker was responsible for establishing and enforcing punishments for infractions of both military and civilian regulations. He was empowered with the authority to perform searches, make seizures of contraband and arrest individuals for violations. Parker was authorized with the command to provide and maintain a jail for prisoners, conduct and manage court-martials (acting as the sergeant-at-arms), administer punishments determined by court-martial and conduct executions and subsequent burials.

Gone were the days of escaping to their Charleston house at 10 Legare Street to enjoy the social season. With the onset of war and Dr. Parker's new role in society, the Parker family undoubtedly spent a lot of time in Georgetown. The family did spend plenty of time at Mansfield during the war, as evidenced in many of Dr. Parker's official correspondence, but probably stayed at their Greenwich Plantation home with greater frequency since it was closer to Dr. Parker's new office in downtown Georgetown.

Parker was able to get away from Georgetown periodically for special events and official business. One such occasion was on April 12, 1861, when he witnessed the historic firing on Fort Sumter in Charleston Harbor that signaled the beginning of the American Civil War. Parker observed the event in the company of fellow Black River rice planter Gus Trapier from the vantage point of Fort Moultrie.[338]

All four of Dr. Parker's sons served during the war. John, his eldest son, honorably joined Kirkland's Cavalry when the war started but was killed in camp near Dumfries, Virginia, on December 23, 1861, when one of his comrade's weapons accidentally discharged. He was twenty-four years old. Frank, the second-eldest son, was aide-de-camp to General Braxton Bragg. He was often mentioned in dispatches for gallantry in action. He quickly rose to the rank of captain and, toward the end of the war, became major and assistant adjutant general. Parker's third son, Arthur Middleton Parker, joined Company A of Hatch's Battalion a month after his sixteenth birthday in 1861 as a private. Later, he served in Parker's Light Horse Artillery, commanded by his relative Captain Edward Parker. At the Battle of Secessionville on June 16, 1862, also known as the First Battle of James Island, Arthur had two horses shot out from under him. James Rose, the youngest of the four Parker boys to serve, joined the Confederate army in the last year of the war with his class of Citadel cadets.[339]

Many of the other local planters served in the Confederate military as well. Like knights of old, these cavaliers of the South Carolina Lowcountry carried the notions of honor and chivalry with them to the contest.

Although a strong Unionist prior to the signing of the Ordinance of Secession, Dr. Francis S. Parker's brother-in-law John Harleston Read II supported his state once South Carolina declared independence and the Confederacy formed. Read, along with Joe Ford and some others, tried

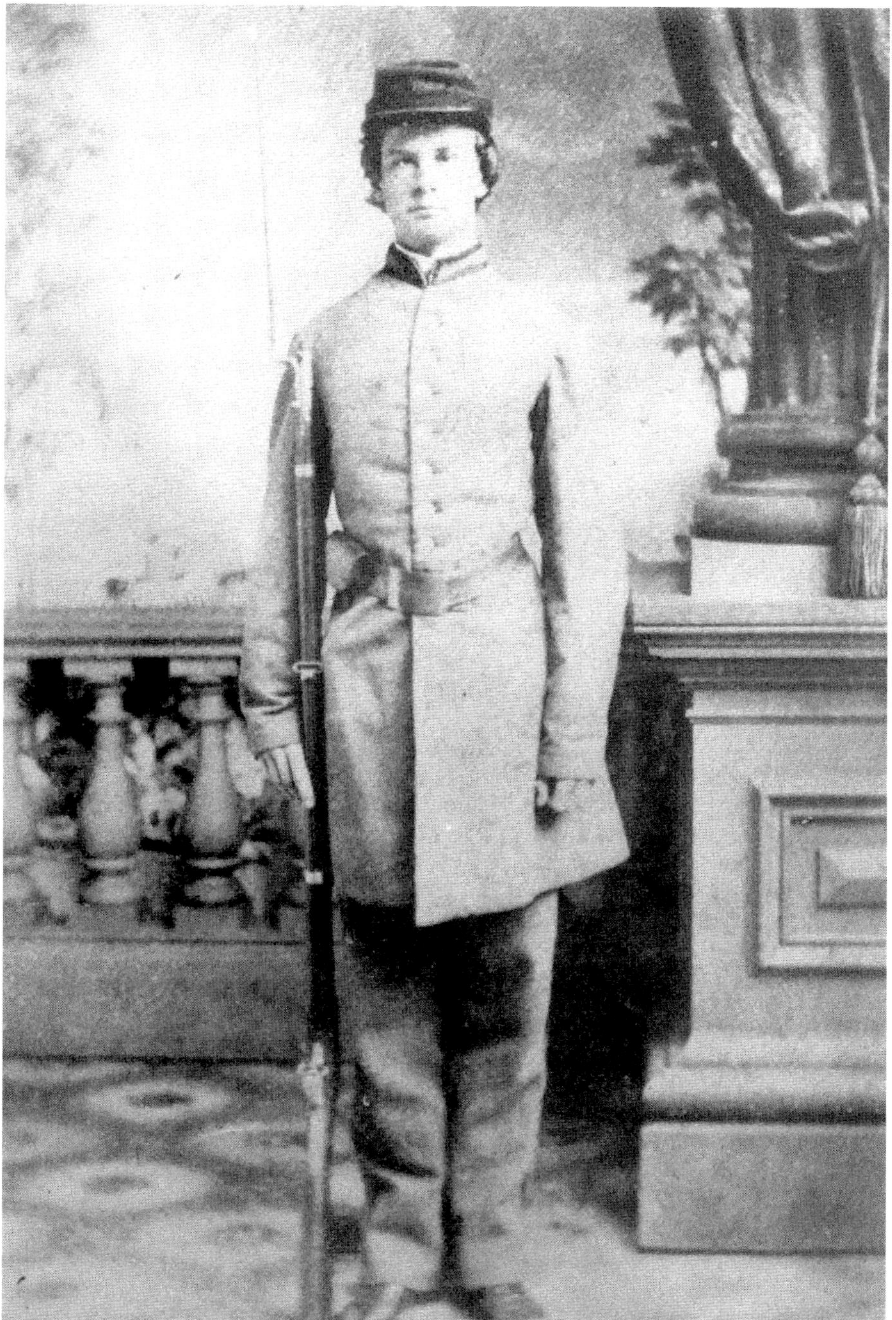

Arthur Middleton Parker, 1845–1912.

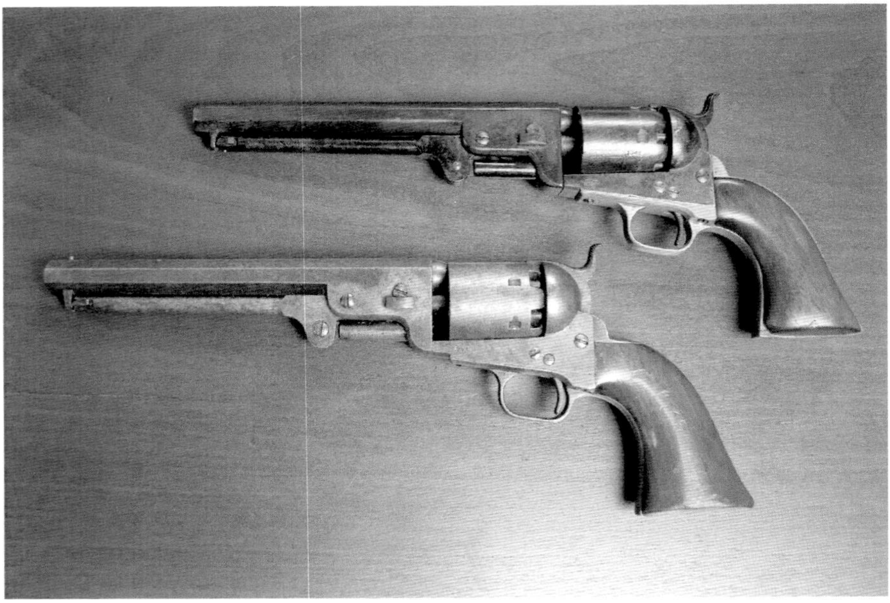

These two pistols were used during the war by Frank Parker and Arthur Middleton Parker.

to rally Carvers Bay and Black Mingo men into joining a company to fight in Wade Hampton's Legion. Only twenty-three men rallied to the call. After two weeks of drilling, the leaders abandoned the cause. Next, Read recruited and equipped a company of Georgetown men at his own expense, including uniforms and weapons. With Read as captain, the company became Company A, Twenty-first Regiment, South Carolina Volunteers, Hagood's Brigade, Army of Northern Virginia. Read was later promoted to major and led his Georgetown troops in many battles in Virginia. His son, John Harleston Read III, volunteered as a private in 1861 and advanced to the rank of captain. He served at Fort Sumter, Cold Harbor, Weldon Railroad, Petersburg and Fort Fisher.[340]

Benjamin Allston, son of Robert F.W. Allston and graduate of West Point, who had served for three years in the U.S. Army before the onset of the contest, formed Marion's Men of Winyah, a cavalry unit that his father financed.[341] The Ward brothers, sons of the late Joshua John Ward of Brook Green, formed Ward's battery of light artillery as captain, first lieutenant and second lieutenant. John H. Tucker, proprietor of Litchfield and Willowbrook Plantations, financed a cavalry unit that featured his son, Dr. John Tucker Jr., as captain. Others joined the

Washington Light Infantry, the Charleston Light Dragoons and the Georgetown Rifle Guards.

Plowden C.J. Weston of Hagley and True Blue Plantations joined the Georgetown Rifle Guards (later renamed Company A of the Tenth South Carolina Regiment) as a private but quickly rose to the rank of captain after feeding, clothing (both summer and winter uniforms) and arming the entire company with English-made Enfield rifles from his personal funds. Weston also donated the use of four of his slaves to serve as a company band (three drummers and a fife) and formed a pioneer corps of four additional slaves to clear the way for the army when it marched through thick underbrush. The Tenth Regiment drilled at Camp Marion and was later attached to General Braxton Bragg's army. Both James Heyward Trapier and Arthur Middleton Manigault, Georgetown planters from the Black and Santee Rivers, respectively, rose to the rank of general in the Confederate army.[312]

Meanwhile, some of the other important men of the era formed a home guard for the defense of Georgetown. Members included Captain W.W. Shakelford, First Lieutenant Samuel Taylor Atkinson (signer of the Ordinance of Seccession) and Second Lieutenant J. Tamplet (part owner of J.J. and P. Tamplet and Company, which owned steam-powered sawmills in Georgetown).

After a meeting in Charleston, provost marshals from throughout the state published an open letter to the citizens of South Carolina in the *Charleston Mercury*. The group created the Aid and Relief Association of Charleston and asked that each judicial district create an association to collect and create supplies to be sent to the soldiers in the war effort. People of the South Carolina Lowcountry sent their donations of supplies and homemade items such as socks, blankets, shirts and drawers to Charleston for distribution to the soldiers through the Quartermaster Department, while the people of the upcountry sent their items to Columbia for distribution through the same means.[313]

By August 27, the ladies of Georgetown had responded by announcing the formation of the the Georgetown Relief Association. Meeting at the Indigo Society Hall, the ladies of Georgetown drew up a constitution and created the organization with Miss H.M. Trapier as president, Mrs. Plowden C.J. Weston as vice-president, Mrs. S.T. Atkinson as corresponding secretary and Miss Anna White as recording secretary and treasurer. Members of the organization included Parker's sister-in-law Mrs. Ester Jane Lance Read, wife of John Harleston Read II, and his stepmother-in-law, Sara Laura Smith Lance, wife of Reverend Maurice Harvey Lance, as well as other plantation mistresses and ladies of the town. Mrs. Parker was not listed as a "manager" at the meeting. She probably had accompanied Dr. Parker to Charleston and was not present in Georgetown for the formation of this organization but was almost certainly involved after the Parkers returned from Charleston.[314]

By December 1861, eight months after the opening volleys of the war and a month after the fall of Port Royal, South Carolina, the U.S. Navy successfully blockaded Georgetown and cut off its trade with Europe. The arrival of the Union ships *Gem of the Sea* and *James Adger* was sufficient in shutting down the port since it was not properly protected by artillery or forts.[315]

On Saturday, January 4, 1862, a day after the South Carolina Senate passed a resolution authorizing provost marshals to use force to compel people to arm in defense of the state, Dr. Parker was again called on to serve his state. Upon recommendations from the coastal districts of Charleston, Georgetown, Beaufort, Colleton and Horry, the state senate passed a resolution to empower certain individuals to be in charge of the removal of slaves and other property from those portions of the state that were in danger of being invaded by the Union. Dr. Francis S. Parker, Robert F.W. Allston and Ralph Izard Middleton were appointed to carry out this task in Georgetown, and Charles Alston Jr., William J. Graham and Francis E. Sessions were called on to perform the same duties in neighboring Horry District.[316]

Soon thereafter, many planters moved their belongings into the interior of the state for the duration of the war. The Union war effort in the Georgetown area did not center on conquering and occupying the town of Georgetown but rather destroying the salt vats and disrupting the rice plantations. By the middle of April 1862, Union gunboats had entered the mouths of both the Santee River and the Winyah Bay. Without meeting resistance in Winyah Bay, the Union navy sailed freely past Georgetown and up her intricate river system, where navy gunboats were able to disrupt the plantation system.[317] The frequency of Union raids continued unabated, and the prevalence of enemy gunboats soon became a common sight on many of Georgetown's rivers.[318] With each consecutive raid, fields were burned, houses looted and fired on by the gunboats, property destroyed and slaves carried away. By the summer of 1862, over seven hundred slaves had been confiscated and deposited on Georgetown's North Island.[319] Abraham Lincoln's Emancipation

Proclamation did not go into effect until January 1, 1863. These slaves were not brought to North Island to be set free but rather to be detained as contraband of war.

On July 21, 1862, the *Charleston Mercury* reported that Union gunboats "come and go whenever they please" and that they had removed between "900 and 1,500 Negroes" to North Island, where they were fed one quart of rough rice a day for their allowance. The newspaper reported, "It is very sickly, and the Negroes are dying daily with a swelling."[350] On another occasion in July 1862, the *Charleston Mercury* reported that the Union ships came to Georgetown one day under a flag of truce and evacuated all citizens of Georgetown who professed to be loyal to the Union.[351]

Without any protection from the Union navy and under order from Dr. Francis S. Parker and other local authorities, most of the planters evacuated the rice-growing district, moving inland away from their plantations. Some of the Waccamaw Planters moved to Conwayborough, up the Waccamaw River or farther into the state, while the majority of the planters on the Pee Dee and Black Rivers moved to Plantersville or inland. The majority of the planters on the Santee and the Sampit Rivers, already living south of Georgetown, moved to Charleston.[352] A few plantation owners (those living the farthest up the rivers) remained at their homes, but most abandoned their homes and crops to rot in the fields. For the remainder of the war, most of Georgetown's rice fields lay fallow.

Captain Plowden C.J. Weston, Company A of the Tenth South Carolina Regiment, was fighting the war nearly one thousand miles from Georgetown. Since the Union forces were looting homes along the lower part of the Waccamaw River, Weston sent his wife, Emily Frances Esdaile Weston, and forty trusted servants to Snow Hill, his wartime refuge in Conwayborough. In the summer of 1862, Mrs. Weston received the Reverend William Wyndham Malet of Hertfordshire, England, at Snow Hill. Malet took careful notes on Conwayborough that give valuable insight into the social and economic conditions during the era.

Malet wrote that the value of rice had dropped from sixteen to eighteen dollars per barrel before the war to eleven and a half dollars by the summer of 1862. By then, planters could not deliver their grain to markets. The price of everything else increased due to the Union blockade. Butter increased to over one dollar per pound, and there was not any ice or tea available. Cotton had increased in price from five cents per yard to forty, and boots that were three dollars per pair before the war increased to thirty and forty dollars per pair. Everything imported to the region was in short supply, including medicines of any kind. The people were forced to use homemade remedies from their gardens and from the woods. People had taken to growing as much as they could in order to survive, including sugar, which Malet noted grew to over twelve feet tall; figs; eggplant; tomatoes; okra; melons; and grapes.

Malet wrote that rice was plentiful but not harvestable in many cases since there was a shortage of labor—most slaves were removed from the district. With Union efforts focused on depriving the locals of food and salt necessary for preserving meat, salt had risen from seven to ten dollars per bushel on the coast, whereas it was fifty cents before the war, and molasses increased from twenty cents to four and five dollars per gallon.[353]

Elizabeth Collins, another English visitor to the Westons' Snow Hill estate during the war, also noted the rate of inflation in Conwayborough:

> *Money seemed to be very plentiful, as is often the case in war times, so that many people would boast of what they had bought at the following prices; calico, eight dollars per yard; homespun, four and five dollars per yard; a box of matches, one dollar (formerly six cents); soap, seven dollars a bar; candles, six and seven dollars per pound; tea (green), twenty dollars per pound; tea (black), ten dollars per pound; wheat flour, one hundred dollars a barrel; beef, two dollars per pound; bacon, two and a half dollars per pound (it rose to ten dollars in 1861); sewing cotton, two dollars a spool; needles, five dollars a packet.*[354]

As seen through the recollections of Reverend Malet and Elizabeth Collins, the economic conditions quickly deteriorated due to the Union blockade, and the quality of life in Georgetown declined. Although the blockade had stopped the local planters from exporting their rice to European markets, well-connected planters such as Dr. Parker and former governor Allston found the Confederate military to be a considerable buyer of agricultural staples and livestock. In the first two years of the war, Parker sold $11,726 worth of rice and Allston $10,078. Other local planters sold beef, corn and fodder to the quartermaster.[355]

Dr. Parker wrote a letter to the Honorable D.F. Jamison (president of the Secession Convention and author of South Carolina's Ordinance of Secession) requesting that he be excused from his state legislature duties so that he could concentrate on his responsibilities in Georgetown.

*Hon. D.F. Jamison:*

*Dear Sir:*

*This district having been abandoned to the enemy, we are subjected in all our rivers to frequent raids by gunboats, attended always with spoliation and destruction of property. The duties of the office I hold* [provost marshal by appointment of the governor] *require constant vigilance and attention. I ask to be excused from attending the Convention* [general assembly] *at the meeting called on the 9th instant.*

*Yours respectfully, your obedient servant,*
*Francis S. Parker*

Dr. Francis S. Parker and Robert F.W. Allston continued to write letters to the War Department encouraging the Confederate army to bring troops to the area to both defend the region from the enemy gunboats and to secure their property, as well as suppress slave uprisings. In June 1862, their requests were answered as General John C. Pemberton ordered Major Emanuel (cavalry) to secure Georgetown. The troops arrived just after harvest in November (it was to be the last substantial harvest until 1866).[356]

At that time (November 1862), Henry A. Middleton wrote a letter to his wife describing the desperate state of conditions in Georgetown. In his dispatch, he recounted a disturbing incident in which Dr. Parker exercised his position against a few captured runaway slaves from Dr. John D. Magill's Richmond Hill Plantation on the Waccamaw River. The slaves ran away but came back to get their wives the next night and were caught. The provost made an example out of the runaways by ordering a free man of color to construct gallows from which to hang them. Parker exercised (in his own words) "wide and extended jurisdiction over the slaves…within this district."[357] Henry Middleton wrote:

*For some months back people have been in numbers leaving different plantations and generally it has been by water—for at night there are few boats afloat. To correct this they are just establishing a regular guard by water. I have given the best of my boats 22 feet long for this purpose—others will do the same. They will play chiefly between Fraser's Point and Dover—Also of the people who went away three*

> *men, returned to the plantation of Dr. McGill* [Dr. John D. Magill was the proprietor of Richmond Hill Plantation in the Georgetown District and Oregon Plantation in the Horry District] *and carried away their wives—the six were taken together making their way to the enemy. The men were tried yesterday by the provost-martial's court—they were sentenced to be hung—today one o'clock was fixed for the execution that no executive clemency might intervene—Dr. Parker came here soon after it was in the gaol yard—a strong military drawn up—there was a crowd—the blacks were encouraged to be present—the effect will not soon be forgotten. As far as I can know Dr. P*[arker] *has acted with great decision and judgment.*[358]

Handling the cases of runaway slaves was a very delicate and complicated matter that received top priority from the provost during the war effort. Men who were not fighting in the war were forced to serve in slave patrols to assist in keeping peace on the homefront. The men in the slave patrol rode the roads and rivers every night and often acted outside of their official authority. They were supposed to apprehend slaves, not punish them without a provost court hearing. On one occasion, one member of the slave patrol shot out the eye of a slave woman who was carrying a baby in her arms as she and a small group attempted to escape to the enemy. On another occasion, a captured slave named Nemo from Dr. John D. Magill's Richmond Hill estate was drawn and quartered by Dr. Magill and Mr. McCuskey for trying to escape.[359] Unlike many of the planters who moved their family and slaves inland during the war, Dr. John D. Magill, who lived far up the Waccamaw River, remained in residence at Richmond Hill Plantation and constructed an earthen fort equipped with a smooth-bore cannon to defend his estate from Union navy patrols.[360]

Writing from Mansfield on November 23, 1863, Parker wrote to William Porcher Miles (former mayor of Charleston and member of the U.S. House of Representatives who was then serving in the Confederate House of Representatives) concerning his observations of the free people of color in Georgetown. Parker stated that this small group of the population was carrying on their lives without sacrifice and was actually profiting during the war effort since all the white men and slaves were busy serving the Confederacy in one way or another. He suggested that the free people of color between the ages of sixteen and fifty be drafted into service to the Confederacy to perform "menial and much of the mechanical service of the army for the war, and at a moderate rate of wages. They might be employed as teamsters, hostlers, musicians, hospital stewards, attendants and nurses, caterers and cooks, shoemakers and cobblers, and general assistant artificers." Parker explained how this act would free many men from services that could be used to fight for the cause of freedom from the United States.[361] Apparently, Miles was not impressed by Dr. Parker's observations and was not moved to act on his suggestion.

Dr. Parker also wrote many letters attempting to encourage superior officers to use the limited resources that the Confederacy had to protect the vital rice-growing region. On November 29, 1863, P.G.T. Beauregard replied to a letter that Parker had written and mailed him on the November 22:

> *Mr. Parker,*
> *Mansfield, near Georgetown, S.C.:*
>
> *Dear Sir:*
>
> *Your letter of the 22nd instant has just been received. I am fully aware of the importance of the country embraced in the Fourth Military District, and of the advantages offered by Winyah Bay as a harbor for certain class of vessels, but I regret to inform you that the limited means at my command in this department do not permit me to dispose of more men or heavy guns for the defense of the Fourth Military District.*
>
> *The enemy at the present seems to have his attention turned entirely toward this place, and the country between here and Savannah, which is also in a comparatively defenseless condition.*
>
> *Any works attempted to be constructed now near the mouth of Winyah Bay, would soon cause him to move in that direction; hence, so long as we have not the means of supporting works there with a strong force, it would be unwise to commence their construction.*
>
> *As regards ammunition, a fair proportion can always be obtained by proper requisitions on the chief of ordinance of this department.*
>
> *Very respectfully, your obedient servant,*
> *P.G.T. Beauregard*[362]

While Dr. Parker worked tirelessly (yet unsuccessfully) to defend and keep order in Georgetown, his eldest surviving son, Francis Parker, assisted General Braxton Bragg, who led troops at Murfreesboro (Stones River), Chattanooga, Chickamauga, Lookout Mountain and Missionary Ridge. Constantly outmanned and undersupplied, the Army of Tennessee rarely saw success, and General Bragg was agreeably relieved on December 2, 1863, and replaced by Joseph E. Johnston. Hearing the news of General Braxton Bragg's resignation after his defeats at Lookout Mountain and Missionary Ridge, Parker wrote to the general on Christmas Eve 1863 in support of General Bragg's service to the Confederacy and in hopes that the Confederate citizens would forgive him for losing battles and history would not forget him. Parker wrote, "A redeemed country will at some future day enroll many names entitled to grateful recollection. Among those to whom a full measure will be due and surely accorded, will be yours."[363]

As the war continued, and more men left Georgetown to serve in the contest, slaves became more rebellious and more difficult to control. Parker called for more and more slave patrol watches at a time when fewer and fewer men were available to help. Eventually, the governor began to call on provost officers to join the regular army and serve the Confederacy. On Christmas Eve 1864, with the end clearly in sight and his world crashing down around him, Dr. Parker wrote a letter to South Carolina governor Andrew Gordon Magrath asking that his officers be exempt from conscription laws since they were needed to help keep law and order in Georgetown. The provost complained that the "constant presence and proximity of the enemy exercise an unhappy influence" over the slaves both in the town and on the plantations. Perhaps trying to qualify his position as still worthy, he went on to explain the punishment of slaves by the Georgetown Court under his tenure as provost marshal. He said, "In the last three years, eleven [slaves] have accordingly been sentenced to suffer the extreme penalty for the law and have been executed, and others have been punished for offenses committed against the peace and good order of the community."[364]

The Union influence over Georgetown and the surrounding countryside was primarily confined to the rivers until February 1865. On February 4, 1865, Parker received an order from R.W. Meminger (assistant adjunct-general) to move all government documents and valuables out of Georgetown to a safe location in the interior of the state. In his letter, Meminger wrote, "The major-general commanding directs that you see that cars necessary for the removal of government stores be vacant in the morning, even should it be necessary to turn out private freight." There was nothing left to be done. The Union forces were clearly ready to land in Georgetown. Under Parker's orders, Georgetown's records were sent to Chesterfield, South Carolina. Parker believed the records would be safe from the invading Union forces; however, the depository was in the path of the conquering enemy, who later burned the documents.[365]

On February 17, General William Tecumseh Sherman took Columbia, and General Quincy Adams Gillmore accepted the surrender of Charleston. Next, Sherman headed north through the interior of South Carolina and into North Carolina, and Gillmore moved on Georgetown. On February 25, 1865, Georgetown officially surrendered. A month and a half later, news arrived that General Robert E. Lee and the Army of Northern Virginia had capitulated to General Ulysses S. Grant at Appomattox and that the Confederacy's president, Jefferson Davis, and the Confederate cabinet were in flight from Richmond.[366] Two weeks later, General Joseph E. Johnston surrendered the Army of Tennessee to Sherman outside Durham, North Carolina. With Johnston's surrender, the war in the East ended. Soon thereafter, soldiers who had been fighting with Lee and Johnston began to straggle home to find Reconstruction at hand.

# *Chapter 9*

# Reconstruction of Georgetown: The Parker Boys in Charge

*Plantation life in the South began with the earliest settlements, towards the end of the seventeenth century, and closed with the outbreak of the war for Southern Independence. In 1860 there perished by the sword of misunderstanding the gentlest, most humane, most chivalric civilization that America has ever known.*[367]
—*Archibald Rutledge*

Although the Southern effort at nation building lay destroyed and Georgetown's plantations had been ravaged, some of the district's planters continued to produce rice. The war disrupted the production and crippled the economy of Georgetown, but the culture would stumble along for another fifty-five years before it completely collapsed.

During the half century following the war, planters faced two very distinct problems in restoring the rice empire. The first pertained to their inability to secure a permanent workforce. The other lay beyond their control. New tools and techniques of mechanized planting, which worked in the prairie states, did not function properly in Georgetown's boggy soil. Planters tried desperately to revive their old rice industry, refusing to accept that the antebellum age of magnificence was forever gone.

On February 25, 1865, after four years of sovereignty, Union soldiers forced Georgetonians to lower the flag of the Confederacy and hoisted the flag of the restored Union over Georgetown. One week later, on March 1, several companies of "colored" occupation troops relieved the Union army in the district. The occupation forces stationed in

Georgetown consisted of the famed 54$^{th}$ Massachusetts Volunteers, five companies of the 102$^{nd}$ U.S. Colored Troops and eight companies of the 32$^{nd}$ U.S. Colored Troops.[368] These groups emancipated the last slaves in the district.

On March 5 and 6, a month and a few days before the fall of the Confederacy in Richmond, the Union army struck the first and most crippling blow to the rice culture. "Colored" Union soldiers and the U.S. Navy sailed up Georgetown's rivers and announced the end of slavery.[369] Jesse Bellflowers, overseer at Chicora Wood Plantation, wrote to his employer and informed her of the quantity of occupation troops in Georgetown: "I understand that some 600 or 800 hundred colored troops arrived in town yesterday and more is expected to arrive in a day or two. Part of them is to go to Kingston (Conwayborough) and the other part to go up the Waccamaw River."[370] The troops informed the planters that they remained responsible for the mortgages that any bank held on their slaves but that the slaves were free to leave.[371] Encouraged by the Union troops, some newly freed men roamed the countryside ravaging plantations and other symbols of the planter's authority, while others remained docile. Elizabeth Blyth Weston recorded the actions of one marauding band on the lower part of the Black River near Georgetown, which destroyed several manor houses, including that of Dr. and Mrs. Parker's Greenwich Plantation:

> *Mrs. Pringle I suppose will write you—I hear she has suffered much having had three visits in one week one at midnight four Negroes and White men that Mary fainted away and she hated to recover her. The Pyatts went at once to Georgetown leaving everything. Toney not having a change of clothes for her infant and I hear has not a servant. Her house was given up to the Negroes at once. Mr. H*[enry] *A. Middleton was ordered to G*[eorge] *T*[own] *his house and buildings burned a faithful Negro got him a change of clothes which he hung on a stick across his shoulder. Mr. Gaillard offered him a horse which he refused I hear he was met by the enemy who took his hat and clothes from him. Dr. Parker went off at once and his place destroyed* [Greenwich Plantation]. *Mr. G* [illegeible] *broke the oath of allegiance and can therefore give us no news. Mrs. G is with him but wants to get away. The Wards (family of Joshua John Ward of Brook Green Plantation) fortunately got off in time to Marlborough. I hear Mr. Wm. Trapier is in prison. The Revnd Mr. Trapier was long*

*in prison and threatened with hanging for poisoning the wine he gave them and his son with him who was made to testify against his father, he is again at liberty but his family have been stripped of everything and the faithful servants who wished to remain with them were carried off by force. They would starve but for Mrs. (Maurice Harvey) Lance who sends them provisions secretly at night. The George Fords were burnt out some time ago and lost almost everything.*[372]

After receiving the news of their emancipation, some slaves simply left the plantations, while others stayed and rioted. On March 18, 1865, Jesse Bellflowers again wrote to his employer informing her that the estate was not a safe place for her to be after she asked if she should come home:

*I do not think so and would advise not to attempt it for the people have behaved very badly and I do think if you had been on the Plantation that you would have been hurt by the people. I have been compelled by them on the place to give up the barn key or to suffer from theare hands. On Sunday the 5th of March two yankeys come up and turned the people loose to distribet the house which they did, taking out everything and then to the smoke house and store room doing the same as in the house and took the plough oxen and kild some of them. The pore mules has been road to death all most after this the people have puld down the mantle pieces broke them to pieceses, taken all the doors and windows, cut the banisters and* [sawed] *out all such as wanted and have taken away the fenceing a round the yard, broke down the old stable and the carpernter shop.*[373]

A few weeks later, on April 2, 1865, Bellflowers wrote another letter explaining a freedman uprising on the Pee Dee River:

*The peaple have divided the cattle...the negroes still go on in puling building to peicies thay have broke in to Brick Church and taken out all the board that was left in it Puld down the club house (the site of the Planters Club on the Pee Dee and also the Winyah and All Saints Agricultural Society) and the two little Barnes up Chapel Creek that belong to Ben the houses that William Put up over Chapel Creek, the Pitman Summer house on the exchange have cut away the banisters in Ditchford house. I hear that thay have made the attempt to Ripe up the floor in Chicora House I have not been down theare in 2 weeks thay*

> *have Kild some of the Goats and some 2 or 3 sheep Molley and Guy has sold off all the poltrey.*[374]

Mrs. Jane Pringle of the White House Plantation wrote a letter on April 1, 1865, describing what had happened at the Chicora Wood manor house:

> *All your furniture has been taken not only by your own negroes but by troops and hordes of blacks who like vultures hung around the plantation here and by their numbers overawed the negroes and got the lion's share of the booty, mostly from Geo town.* [Your] *woman Lavinia claimed and took all the furniture to Plantersville house. As to the cattle, sheep and hogs these were I know ordered by Col. Brown himself to be shared among the negroes on each absentee place.*

Later in her letter, Pringle wrote, "Had I had poison in the house on those two days of the great orgy I should probably have given it to my daughter and taken it myself, for I despaired."[375]

During the months of March and April 1865, corruption and crime reached epidemic levels in the Georgetown countryside. Soldiers and bands of freedmen looted the manor homes and carried off the planting families' valuables. One plantation mistress wrote to Colonel Brown of the Union army explaining the actions of the freedmen and asked to be escorted home to her plantation. When she and others of her social set returned home, they quickly realized the magnitude of change that had taken place. The proprietor complained that her manor home had "been robbed of every article of furniture and much defaced and injured also all my provisions of meat, lard, coffee and tea taken. That we are left without a bed or blanket and that all these necessary things are distributed among the Negroes."[376]

Richard Lathers recalled the destruction and theft of priceless pieces of art that the people of Georgetown owned prior to the war that was lost during Reconstruction: "Before the destruction of art collections in the South by the Civil War, the Georgetown District possessed many of the rarest and best specimens of Reynolds, Copley, West, Allston, Stuart, Sully, Lely and Gainsborough."[377]

In Georgetown and throughout the countryside, the presence of the "colored" occupation troops gave the ex-slaves a strong sense of impunity. Joel Williamson pointed out in his book *After Slavery* that it was always more difficult to maintain law and order in areas occupied by colored troops. The freedmen in these situations often did what they pleased and disregarded laws because they did not expect to be punished for any of their actions. Mrs. Pringle wrote a letter explaining the actions of one group. She explained that "the freedmen were celebrating their emancipation. The blacks had a large dinner party on Dr. Sparkman's place last Sunday and got drunk and have invitations out for a picnic."[378]

The use of black occupation troops was not confined to Georgetown, nor was their presence rare in the Lowcountry. In fact, the Union army stationed colored troops in every district of the Lowcountry from Georgetown, South Carolina, to Savannah, Georgia.[379]

After losing their slaves, the planters knew their only chance to rebuild their fortunes rested with continued ownership of their lands. On March 3, 1865, the Freedmen's Bureau enacted General William Tecumseh Sherman's field orders, which gave each freed slave forty acres and a mule. The Union army distributed confiscated land south of Charleston to the freedmen.[380] This unfair land confiscation and distribution policy led Georgetown's planters to fear that the same fate might befall their property. Many of the planters who had earlier moved inland to escape the war moved back to their plantations to claim their holdings.

On April 25, under orders, some of the planters of the Georgetown and Charleston Districts took oaths of allegiance to the United States. Afterward, the Union army instructed the planters to assemble the freedmen and draw up formal labor contracts. Planters also had to provide for the freedmen as they had before emancipation and share the harvest with them.[381] The occupation forces warned the planters that if they decided not to grow crops on their land, they would still be held responsible for providing for their own and possibly other ex-slaves on their land, including feeding and housing them.[382]

Rice was planted shortly afterward, but due to the shortage of laborers and the fact that the fields had lain fallow for four years in many places, the yield was very small. The Freedmen's Bureau planted and managed several plantations, most of which were on the Waccamaw River. The ex-slaves worked hard until after the crop was harvested. Afterward, however, some refused to repair broken trunks or work at ditching.

On October 20, 1865, Dr. Francis S. Parker, Benjamin Allston, Joseph W. Ford, Paul Fitzsimons and S.S. Fraser asked Lieutenant Colonel A. Willard, the local commanding officer, to force the freedmen to work at ditching. Their petition stated that poor drainage would be a health hazard to the entire community. After deliberating over the crisis, the U.S. Army used legal force to make the freedmen perform the tasks that they usually performed after harvest.[383]

Dr. Francis S. Parker attempted to resume his place in society after the war by returning to politics. In 1865, he served as a magistrate of Georgetown, along with John Harleston Read II and eight others. He also served as a commissioner of public buildings, free schools and roads.[384]

Although the majority of the planters were unsuccessful planting rice with the new labor system, many attempted to take care of their tenants' needs. Daddy Ancrum, a freedman, remembered his employer as a fair individual who fed his family during the summer and fall of 1865 before the crop was harvested. Ancrum stated:

> *Mauss Ben he done puty by me; I had nine head to feed, an Mauss Ben say he feed them all fo' my wuck, so Mauss Ben feed my family fu' dat year, and feed dem well, an' we mek fine crop o' rice. The fust contract was you fu'nish land and seed and animals an' get two-thirds; I fu'nish wuck and get one-third. Every day I didn't wuck was deduct' from my share.*[385]

The war and the years immediately following had an enormous psychological effect on the Georgetown planters and southerners in general. White southern males had been beaten militarily, and the flower of their generation was either killed or maimed. With Georgetown under Union occupation, planters had a very difficult time getting laborers to follow the labor contracts they signed. To add to the problems, in some cases, former slaves did not hesitate to break into rice barns and smoke houses and steal food from the planters as a form of overdue payment.

Money was very scarce, and since there had been so much destruction immediately following the war, and planters had problems convincing freedmen to work the 1865 crop, the winter of 1865 was difficult for everyone, with a severe shortage of supplies and food. In December of that year, Adele Petigru Allston wrote to her daughter Adele Vander Horst describing the condition of the former slaves. December was the time when most planters traditionally handed out winter rations and clothing allotments. Without any allotments to give, Mrs. Allston wrote:

> *The Negroes are as unsatisfactory as possible, the prospect ahead discouraging. The poor nigs have certainly deteriorated very much in*

> *appearance and manners in the last year or eight months. They are without shoes or clothes and* [though] *I spent plenty of shoes and swine of various kinds nothing that looks like the comfort of former days. Our prospect for meat is poor indeed.*[386]

The winter of 1865–66 was a tough time at Mansfield both economically and mentally. Longtime friend of the Parkers and local military hero Brigadier General James Heyward Trapier, a Georgetown native and West Point graduate of 1838, died at Mansfield and was buried on December 22, 1865. Alexander Glennie performed the service for the general's family. Trapier had been assigned to the Fourth South Carolina District, with his headquarters in Georgetown, since November 1862. Trapier and Dr. Francis S. Parker worked together defending their part of the Confederacy from both Union invasions and local insurrections.[387]

With true abject poverty setting in among the freedmen and desperation among the planters, violence often took place on both sides when planters attempted to restore their authority at war's end, and Mansfield Plantation was not without its episodes. On February 8, 1866, the local magistrate fined Dr. Parker fifty dollars for assault and battery against a freedman named Edward Smith.[388]

When Brigadier General Trapier's widow asked Dr. Parker to help her to start planting again at her Keithfield Plantation, Dr. Parker acquiesced. With the aid of the Freedmen's Bureau, Francis S. Parker drew up contracts with forty-six freedmen and women. Of those listed on the labor contract, two were foremen and two carpenters, including Parker's old driver Dick (previously the driver at Mansfield Plantation).[389] The conditions of the contract were that the freedmen would be paid one half of the rice, corn, peas and potato crops (after one-fifth was deducted for plantation expenses). The Parkers furnished implements, wagons and mules, while the freedmen were to keep the fences and ditches in good repair.[390]

It seems that at one point, Parker was unhappy that the freedmen were not on task and the fields were not prepared for planting (apparently, the freedmen refused to adhere to all of the conditions agreed upon in the contract after work began). In March 1866, Parker sent his son Frank and a former driver named Dennis Hazel, whom Parker wanted to install as overseer, to Keithfield Plantation. The freedmen appointed their own foreman, a man named Abram. Abram quit working on March 31 and called for all the other freedmen in the fields to stop working as well. Parker threatened Abram, saying that he would report him to the authorities and that they would "break him." After further exchange, Abram appeared to be ready to go back to work in the field. Suddenly, Abram called the freedmen in from the fields, and the mob armed themselves with axes, hatchets, hoes and poles and threatened Parker and Mr. Hazel. The two men left but called for troops to meet them later to put down the insurrection. When only two soldiers came to his call, the freedmen were not intimidated by Parker. The freedmen assaulted Parker and his company with their tools, threw bricks and stones at them, threatened them with poles and even took the gun away from one of the soldiers.

The freedwomen in the group had a particular hatred for Hazel. Two women, Sukey and Becky, began to hit him with heavy clubs. The soldiers attempted to stop the beating, but then freedmen joined in the scuffle. Parker pleaded with the women to stop hitting Hazel and promised to replace Hazel, but by this time the crowd was calling for the women to kill Hazel. The one soldier who still had his gun threatened to shoot into the crowd, but Parker begged the soldier not to shoot the freedmen. Instead, Parker yelled out to the women, who then turned their anger toward him. Sukey took a hickory stick and beat Parker over the head. Then Becky joined in on the attack on Parker, who later reported that she hit him in the head, striking him "very heavily over my right eye with a club—the blow bringing instantly, and making me stagger with blindness." Bloodied, without weapons for defense and fearing for their lives, Frank Parker, Hazel and the soldiers jumped into the river to swim to safety as the freedmen threw projectiles at them.[391]

Frustrated at the loss of control over the local labor force and also because Union troops were not willing to help the planters, Frank relocated to Charleston to run the family's Hayes Plantation.[392] Concerning the episode, Dr. Parker said to Colonel Smith of the Union troops, "No white man could control the Negroes now that they were free."[393]

Shortly before his death in September 1866, Dr. Parker's brother-in-law John Harleston Read II made similar contracts for hired laborers at Reverend Maurice Harvey Lance's Mauricena Plantation, as well as his own Harmony and Maryville Plantations.[394] Frustrated, Read wrote, "No negro shall stay on any place I have the management of without being entirely subject to every order I may give." [395]

Like Parker, Read experienced threats against his manager and witnessed similar labor problems on his estates. Besides the rebellions on his own estates, he wrote of similar problems in the surrounding countryside: "Every negro's head is now so full of politics that he can't work. I hear of one entire gang of the men from one plantation coming to Georgetown to attend the Union League meeting and returning next morning...most of them so drunk as not to be able to go into the field at all."[396]

In 1866, Dr. Parker and his family escaped the "sickly season" (summer months) by retreating to Stateburg. While in Stateburg, the Parkers occupied the Episcopal Rectory known as Woodville. That October 11, Woodville was the site of a double wedding for the two eldest Parker girls. Anna Mary married Benjamin F. Huger of Charleston, and Julia Rose married John H. Burgess of Enfield Plantation in Stateburg. When the summer ended, Dr. Parker and his family moved back to Mansfield to face the depressed rice culture and the social and political problems that gripped Georgetown.[397]

One Georgetonian later described the root of social issues between white and black Georgetonians in his memoirs:

> *The Negroes were not made free by a wise process of gradual emancipation, as had been done in the Northern states; but rather ties which bound them to their masters were rudely burst by war and a sweeping constitutional amendment. Thus the masses of Negroes, untrained, improvident, ignorant, and shiftless, were suddenly thrown on their own resources.*[398]

With violence all around them, planters felt helpless and feared that the freedmen might rise up and take revenge for their servitude. Dr. James Ritchie Sparkman wrote to Benjamin Allston informing him of a rumored uprising of which "three of the most respectable citizens" informed him. Sparkman wrote, "White male adults and children were to be massacred, on a given night between this and Jan[uar]y. All the aged White women were also to be destroyed, but females between certain ages, to be reserved for servile and licentious purposes."[399]

Though unwarranted in this respect, the planters had many fears and trials to face. The following years brought crop failures, and by 1867, many planters faced bankruptcy. The struggle to retain authority over the workforce, exorbitant interest rates and the lack of capital (due to crop failures) caused problems, and it appeared that the rice industry could not be revived. In 1867, planters cultivated only twelve thousand acres of rice land compared to forty-six thousand acres in 1860, one year before the war.[100]

In the summer of 1867, Dr. Francis S. Parker desperately attempted to make Mansfield profitable and rebuild his family fortune and prestige in South Carolina. With unfulfilled labor contracts, depressed real estate values and increased costs of labor, his personal fortune was lost. The loss of his son John to war and Frank's constant feuding with the local labor force impelled him to move to Charleston. Parker and his entire world had crumbled since that optimistic day in December 1860 when he signed the Ordinance of Secession. On September 12, 1867, he died at Pineland, his summer home near Georgetown—it's said of a broken heart. Dr. Francis S. Parker was buried the following day, with Reverend Alexander Glennie performing the services. Dr. Parker was fifty-two years old.[101] In 1868, Mary Lance Parker, now a widow, signed over the deed to Mansfield Plantation to her sons: Arthur Middleton, James Rose and Rutledge.[102]

Concerning the status of Georgetown's social and economic structure, Francis Withers Pickens wrote in 1867: "All society stands now like a cone on its apex, with base up. The least jostle is liable to make a total overthrow of the whole structure. Such a state of things cannot stand, for it never has in any age or in any country as a permanent structure. I think, by next May we will begin to see the light, and we will at least begin to know what to expect, and to try and adapt ourselves to it."[103]

Politically, great changes took place in the South as a result of the Reconstruction Acts of 1867. Possibly the most upheaval came when ex-slaves gained the right to vote and hold office, privileges now withheld from white men of prewar prominence. The acts also divided the old Confederacy into five occupied zones (except Tennessee, which was quickly readmitted to the United States) and established military rule. Concerning the state of affairs, John Harleston Read III sarcastically wrote the following to a friend: "What a delightful state of things now exist in the unfortunate South! What glorious liberty we enjoy, where such a man as (Daniel) Sickles, nicknamed 'Devil Dan,' is made absolute Lord and master over two states and dictates such laws as suit his fancy."[104]

The military empowered the newly elected Congress to write a new state constitution and pass the Habeas Corpus Act, which greatly expanded the right of citizens to remove cases to federal court.[105] The Reconstruction Acts transferred political power throughout the South from former Confederates to freedmen and their Republican allies. It was during this time that Georgetown planters began to seriously consider an alternative to their traditional African American labor force.

After more and more of the freedmen left the district in search of work elsewhere, the Georgetown Agricultural and Mechanical Club endorsed a resolution to import Chinese laborers to work the rice fields. Obviously playing into the racial unrest of the area, William M. Lawton, the chairman of the committee on Chinese immigration, wrote a letter to Dr. James Ritchie Sparkman expressing his intentions to "bring in a race of people of a higher type of civilization, ingenuity, and indefatigable industry, to substitute that description, which were brought & sold to our Forefathers, through Puritan, Yankee, & British cupidity, and then destroyed by their hypocrisy & folly."[106] Although unsuccessful at convincing Sparkman and others to accept Chinese substitute workers, the plan and Lawton's proposal are of considerable interest.

William M. Lawton was not the only outsider with plans for reviving Georgetown's rice industry. Oliver Hudson Kelly, the founder of the Patrons of Husbandry, wrote a letter to Benjamin Allston urging him to help set up a chapter of the organization in Georgetown. Kelly told Allston he could encourage immigration to the South using the same techniques used to attract immigrants to the West. Kelly claimed, "Our association will give as much attention to one portion of the country as to another." The Patrons of Husbandry never formed a chapter in Georgetown, but theirs was not the last attempt in procuring immigration for the district.[107]

At Mansfield, the seat of the Parker family's depressed rice empire, the contract labor conditions were slightly different in 1868. Twenty-two laborers (including Porteus Mitchell, the foreman) signed into an annual contract that would last from January to January. The laborers agreed to obey all orders and to observe all rules the employer made for the plantation. Mansfield's payment conditions were more favorable than the 1866 conditions. The new agreement provided laborers with one-third of the rice crop and one bushel of corn per month to each laborer and allowed each laborer the privilege of raising his own garden and corn.[108]

Ex-slaves who remained in the Lowcountry vowed to make a difference in the 1868 elections. South Carolina sent a predominance of African Americans to Congress. African Americans won U.S. Senate and House of Representatives seats, the state senate and House seats as well as local positions. The same political revolution took place in the rest of the state. Georgetown's one state senate seat went to Joseph H. Rainey, an African American who was a free person of color before the war and whose family actually owned slaves.[109] This pattern continued throughout Reconstruction, and Rainey stayed in elected offices into the early 1900s.[110] This group, along with some northern Republican influences, wrote the 1868 South Carolina State Constitution, which guaranteed civil and voting rights to African American males and state-funded public education.

In his book *The Prostrate State*, James Pike called the Reconstruction era the "tragic era." This period of planter "Negrophobia" was chronicled by Pike as being a Yankee-dominated era with corrupt government, increased taxes and unqualified politicians in office. Approximately 68 percent of the members of the new Constitution Committee were African Americans, some freedmen and others born free.

With ex-slaves and former free persons of color in control of the local government, the Freedmen's Bureau (an organization responsible for helping smooth the way for freedmen to make the transition from slave to citizen) began to end its work in Georgetown. The Freedman's Bureau (the first government welfare program in American history) provided freedmen with education, clothing, food and medical facilities. After 1869, however, federal funding for the Freedman's Bureau waned (local bureaucracy administered those programs, along with efforts to further develop a wage-earning labor system).[111] During the same year, the U.S. occupation forces withdrew from the county, and the freedmen created a local militia to ensure they maintained their authority.[112]

Outnumbered and feeling unjustly treated with loss of control over politics and with labor and race problems continuing, some planters followed the lead of other reactionaries throughout the South and created a chapter of the Ku Klux Klan. The first Ku Klux Klan in American history, a paramilitary organization that was formed to scare freedmen, carpetbaggers and scalawags into voting for the Democratic Party, was short-lived. Federal troops waged war against the Klan until it was driven underground in 1871 after the passage of the Ku Klux Klan Acts. Under the acts, President Ulysses S. Grant declared marshal law in nine South Carolina counties as well as other counties throughout the South.

While the freedmen thought the Klan was brutal and unnecessary, the white people of Georgetown considered the Klan a necessary organization. According to Richard Lathers, the Georgetown chapter of the KKK formed because the local magistrates refused to protect the property rights of the planters and white citizens. Lathers claimed the freedmen "frequently stole cattle, hogs and portions of crops while they waited for hand-outs from the Freedmen's Bureau, and later the town." He justified the Klan by stating, "There is not a case on record of hostility to the Union, or any evidence of robbery or desire for personal profit in the Ku-Klux outrages: they were simply crimes of resentment against other crimes."[413]

Hagar Brown, an ex-slave at the Oaks Plantation, remembered having problems with the KKK during Reconstruction. When asked about the Klan in 1937, Brown stated:

> *Ku Klux...Have you run way, you go church. Going to come in to ketch you or do any mischievous thing—come carry you place they going beat you—in suit of white. Old white man to Wilderness Plantation. Parish old man name...Come to house, ain't crack. Come right in suit of white. Drag him out—right to Woodstock there where Mr. Dan get shoot...Come in. Grab you and go. Put a beaten on you till you can't see. Know they got a good grub to lick you wid. They git done you can't sit down. Ain't going carry you just for play with.*[414]

No membership role of the Ku Klux Klan exists. Records prove that the Parker boys did their best to maintain the social order of the Old South. In 1869, Arthur Middleton and James Rose Parker joined the local militia populated by white citizens who were concerned with retaining the rights of the old guard and the Democratic Party.[415]

In 1869, the Parker family witnessed two more marriages, but in 1870, they experienced the loss of Emily Rutledge Parker, who died on Christmas Day at the age of twenty-three and was buried on December 27.

Frank married Eliza Battle Troost in Charleston in May 1869. Mary wrote her son a touching and affectionate letter on May 4:

> *Charleston, May 4th, 1869*
> *Thursday*
>
> *Your wedding day! And a fond word of love and tenderness must go to you my darling son, although from the effects of the voyage I am by no means very bright and your grandfather is in his bed very much the worse for the trip—I say tomorrow he will be all right, however. I find Anna looking well and Frankie almost too fat.*

*We are all thinking continually of you dear Frank, and may God's blessing rest upon you and yours my son. Arthur must be with you eve this time, and yesterday as soon as the boat arrived, 2 o'clock, I posted a letter for him from Emma.*

*Our affectionate love to dear Lillie, and quite a disappointment it is to not be that Anna is so unable to urge his immediate presence in the city.*

*Your affectionate Mother,*
*Mary T. Parker*

*James' letter and invitation arrived Sunday morning in Georgetown*[116]

Soon after marrying Eliza (Lillie) Battle Troost, Frank sold the Parker family's Hayes Plantation to Francis Holmes (who renamed the estate Ingleside) and moved to Mobile, Alabama, with his bride. He did not return to Mansfield Plantation for another fifty years.

Just six months after Frank's wedding, on November 24, 1869, Reverend Alexander Glennie married Arthur Middleton and Emma Izard Middleton at her home at Windsor Plantation (Wee Haw) on the Black River.[117] Arthur Middleton and Emma Izard Parker eventually had four children of their own. Mary Taylor (named for Arthur Middleton's mother) was born on November 29, 1870, and baptized at Prince George Winyah Church on March 15, 1871. The couple named their first son Francis S. (in remembrance of Arthur Middleton's father) on September 21, 1872. James R. Parker sponsored little Francis's baptism on January 19, 1873. The couple named their second son Arthur Middleton after his father. Little Arthur was born in August 1877 and sponsored for baptism by his uncle Rutledge Parker on October 28, 1877. The couple named their fourth and last-born child Ralph Izard Middleton. The child's uncle Daniel Parker and others sponsored his baptism on October 26, 1879.[118]

Although it still led in production during the second half of Reconstruction, South Carolina's dominance in the rice industry quickly waned.[119] There were many problems with the rice industry; however, Georgetown rice, referred to as "northern rice" because it was grown north of Charleston, was still considered the finest-quality rice produced in the United States. It sold in the Charleston market for seven or eight cents per pound, an average that was just double its prewar value.[120]

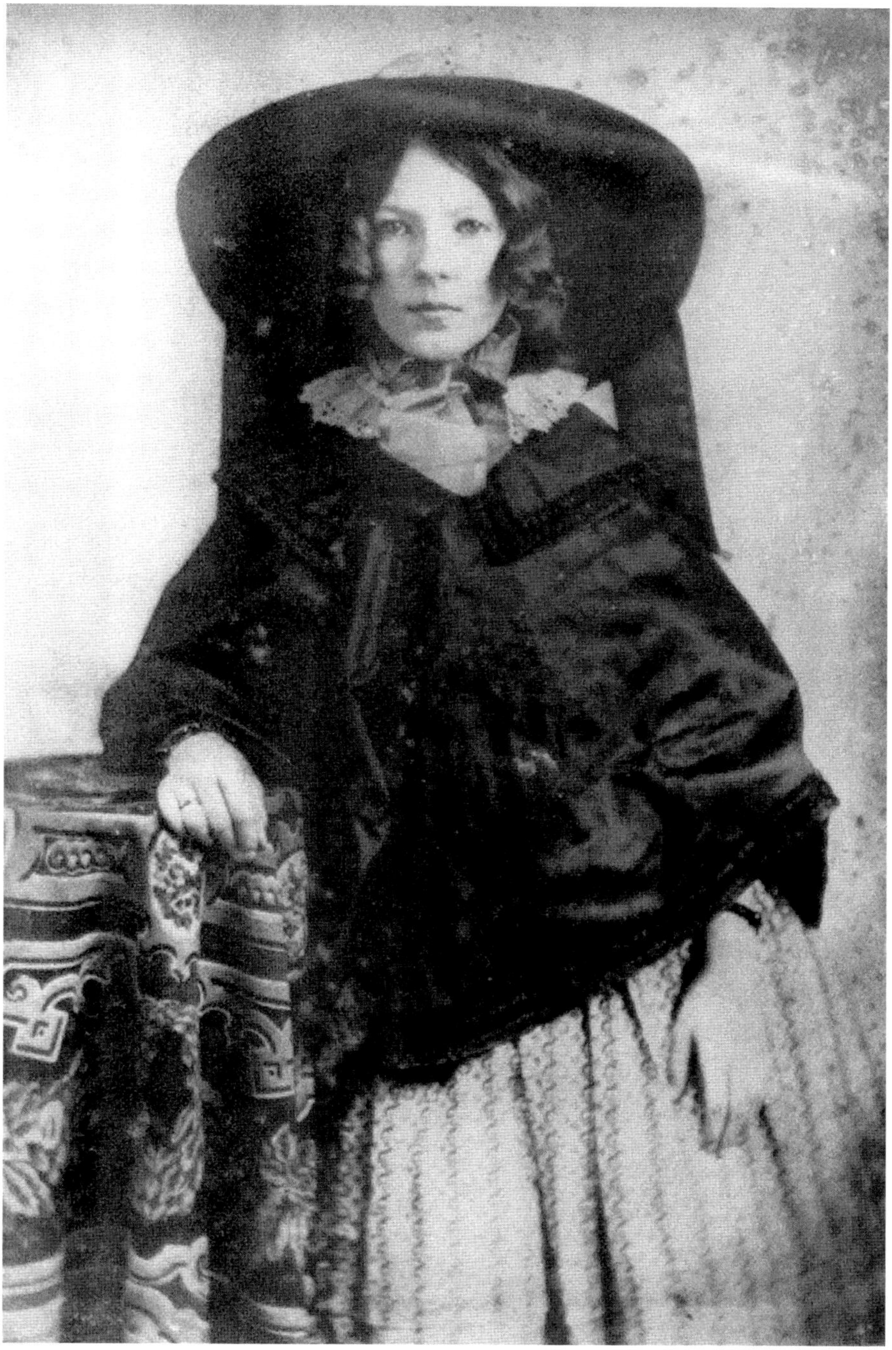

Emma Izard Parker, October 29, 1847, died May 21, 1937.

In an attempt to keep their workers happy and productive, many of the plantation owners changed their payment policies. Some planters began to pay their workers weekly, and some increased their salaries to an average of six dollars a month, plus rations. Those workers who did not partake in receiving rations were usually paid about thirteen dollars a month.[121] Due to the lack of currency, some planters issued promissory notes, which looked like currency, and promised to redeem them in the future.[122]

By 1870, planters had harvested fifteen thousand acres of rice in Georgetown that produced an average yield of 540 pounds per acre. At that time, South Carolina still produced the vast majority of rice grown in the United States, dwarfing the production of both Georgia and Louisiana combined. But South Carolina's lead would soon be eclipsed by these states. Although South Carolinians worked fifteen thousand acres, they had once worked almost fifty thousand. Along with the decrease in number of acres in production, the value of farming implements and machinery decreased from an all-time high in 1860 of $616,774 to an all-time low of $15,862 in 1870. At that time, too, the cash value of all farms in the county had dropped from an all-time high of $5,818,690 in 1860, just prior to the war, to $678,089 in 1870.[123]

By 1870, the population of Georgetown had decreased significantly as well. In 1860, on the eve of the war with the United States, Georgetown had 3,013 whites, 183 free persons of color and 18,109 slaves. In 1870, ten years later, the white population had dropped to 2,773, and due to the freedmen leaving the county, the black population had dropped to 13,388. It was cities such as Charleston that witnessed the largest increases in the black population during the 1860s and 1870s.[124] As a whole, the state's white population dropped from 291,300 in 1860 to 289,667 in 1870, while the black population increased from 412,300 in 1860 to 415,937 in 1870.[125]

At Mansfield Plantation, Arthur Middleton Parker was head of the household. He worked 240 improved acres. At that time, the value of his farm had depreciated to $20,000, and the value of his farming implements was $400. He must have had a considerable number of laborers. In 1869, he paid $3,800 in wages a year (but the estate's rice production was a mere 144,000 pounds, a tenfold depreciation since 1860).

With a combined livestock value of $1,340, Parker had three horses, four mules, ten milch cows, ten oxen and twenty sheep. Even with the great quantity of livestock, including work oxen, rice production dwindled.

In 1870, Reverend Maurice Harvey Lance died at seventy-seven years and five months of age and was buried on November 18. Lance's grandson Arthur Middleton Parker purchased his Mauricena Plantation, where he and his wife, Emma, lived until they moved back to Mansfield in 1885.[426]

Lance had experienced depreciation in his production and the value of plantation over the last decade as well. While working 200 improved acres and owning an additional 1,250 unimproved acres, he produced 150,000 pounds of rice. The value of his farming implements was $1,000, and the value of his farm was $20,000. Lance spent $4,000 on labor wages. His combined livestock was valued at $800. Among his livestock, Lance owned seven mules, three milch cows, four oxen and twenty-five sheep. Interestingly enough, he did not own a single horse.[427]

In 1873, the planters and merchants of Georgetown combined their efforts in an attempt to restore the rice culture. They formed the South Carolina Rice Plantation's Trust. The group's main goal was to attract foreign aid as investors. The trust printed a pamphlet that included statistics on rice and estimates of cost and production for standard-sized plantations. It addressed several other helpful and important topics concerning the rice culture; however, it misinformed the reader concerning the freedmen population. Many of the best rice workers moved on to other enterprises or moved away from the area. The pamphlet claimed that a freeholder would have no difficulty "procuring an ample supply of efficient and instructed labor from the negroes upon the spot. These people, warmly attached to the localities where they were born and reared, are peaceful, orderly and willing to work when honestly and fairly treated."[428] Reading much like a New South booster, the author of the pamphlet stated, "Let the capitalists come and offer a fair day's wage for a fair day's work, and abundant labour, skilled in agricultural work, will respond to his call."[429]

The trust failed to draw foreign investors to Georgetown, however, and the economy continued to decline. Some potential investors surveyed large tracts of land but backed out because the soft, muddy soil prevented the use of machinery in the area.[430]

While new technology developed and planters began to use new machinery in the Southwest, Georgetown held on to the "task system."[431] By 1876, Louisiana began to reap more rice than South Carolina by a considerable margin, and Georgia's production steadily increased as well. Meanwhile, South Carolina's production of the crop remained stagnant.[432]

Some of the planters realized that they were not going to be able to persuade the freedmen to work as they did before the war. As a solution to the race and labor problems, many planters began to agree with the "Back to Africa" campaign. Jacob Motte Alston, the former owner of Woodbourne Plantation on the Waccamaw River, realized the magnitude of the problems between the races but did not believe in sending the freedmen to Liberia. He agreed with the Back to Africa campaign but felt that Liberia was not the correct place to send the freed slaves. Alston offered his diary a humane compromise to the question:

> *To lure them on to Liberia is unwise, unjust, and cruel. Up the Congo River to some region near Victoria Niganza, where a climate similar to that of the lowcountries of the Southern States, where the soil is fertile and where there are thousands of miles of navigable water courses, is the destined home of the negro. There he will be free from the aggressive peril of the white man, and there prove himself a man.*[133]

Near the end of Reconstruction, Georgetonians revived one element of antebellum social and political life. Whites began to organize clubs; however, these new organizations bore little resemblance to the societies of the wealthy planters. On August 20, 1874, in desperation—and possibly as a rebuttal to the black militia and the fact that the Ku Klux Klan had been made illegal—white citizens formed the Georgetown Rifle Guards (named after the famous Company A of the Tenth South Carolina Regiment). At its first meeting, held at the Winyah Indigo Society Hall, the organization claimed that it had formed "a club for self defense" whose belief was "the promotion of social intercourse and the enjoyment of the members by means of target shooting and such other amusement as they may determine."[134] At this time, Georgetown also had the Santee Gun Club.

On October 31, 1876, longtime friend of the Parker family and former Confederate general Wade Hampton III spoke in Georgetown during his bid for governor. At this time, both the Red Shirts (followers of Wade Hampton) and black militias drilled in South Carolina and even clashed in North Augusta. In this heated political and social environment, Hampton chose not to speak in the city but rather at the remains of a former Parker estate, Greenwich Plantation, which had been burned by a band of marauding freedmen at the close of the war. With the plantation's chimney remains draped in a palmetto flag, General Hampton urged the

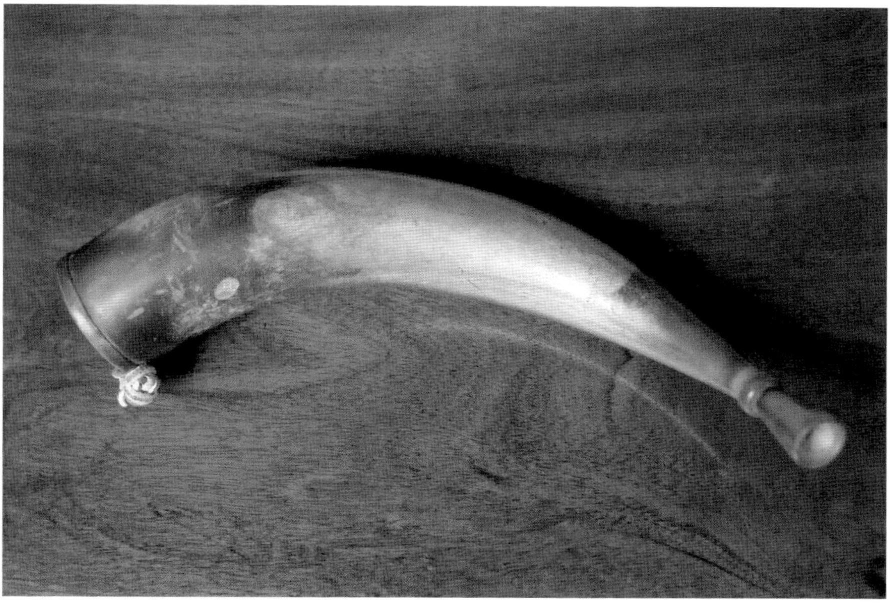

This hunting horn belonged to Wade Hampton. It was used to call deer hounds. He gave it to Dr. Francis S. Parker to signify a political and social alliance between the Hamptons and Parkers.

freedmen to move to the front of the crowd so that they could hear his plan. He spoke of interracial unity in an attempt to revive the economy and promised the former slaves that he would protect their rights and restore peace and prosperity.[135]

The results of the election were not determined until April 1877 due to a compromise with the Republican Party. The election of Wade Hampton and the return of the Democratic Party marked the end of Reconstruction in South Carolina. With home rule restored and the white population "redeemed," the planters looked forward to a revived economy. Contrary to their expectations, the end of Reconstruction had a paradoxical effect on the South Carolina rice industry. New competition from other ex-Confederate states forced the continued decline of the culture. In addition to the effects of the failed revolution and labor problems of the Reconstruction era, mechanized competition proved to be the crippling blow to the industry, leading to its utter demise.

# Chapter 10

# The New South: Arthur Middleton Parker

*The New South was, accordingly, a history of continual redefinition and renegotiation, of unintended and unanticipated consequences, of unresolved tensions. People experienced conflict within their own hearts and minds; classes, races, and partisans clashed. The New South was an anxious place, filled with longing and resentment, for people had been dislodged from older bases of identity and found no new ones ready at hand.*
*—Edward L. Ayers*[436]

The intentions of Reconstruction (1865–77) were to rebuild the South to become a mirror image of the North by uprooting political, social and economic traditions and rebuilding the infrastructure. Reconstruction did change the South from being a stronghold of the Democratic Party to being a Republican Party–dominated region by granting freedmen the right to vote.

In 1880, the two political parties eventually agreed to share power in Georgetown. This fusion plan, also known as the "Georgetown Plan," provided for uncontested control over local government. In this plan, the Democrats (whites) were given control over the sheriff's department, clerk of court, coroner, two county representatives and one representative. The Republicans (blacks, scalawags and carpetbaggers) gained uncontested control over the senate seat, one representative, probate judge, school commissioner and one county representative. Georgetown senator Bruce H. Williams hoped the plan would soon become a statewide one because it gave "a fair and equal distribution of the offices without regard to race or party affiliations."[137]

Instead of solving race problems, Reconstruction failed to create a lasting social revolution and sowed the seeds of legally enforced racial segregation throughout the entire United States.

Although segregation was not regulated by law until the reading of the Supreme Court decision in the *Plessey v. Ferguson* case, it was carried out by both sides during Reconstruction. The University of South Carolina, the former finishing school for the planter elite in South Carolina, was almost entirely populated by freedmen after the war, and nearly all of the scholarships went to freedmen as well.[138] George C. Rogers, in his book *The History of Georgetown County*, points to the fact that blacks and whites began to attend different churches during this time period as a clear indicator that African Americans wanted segregation. At this time in Georgetown, African Americans had two churches, the Methodist and the Baptist, whereas the white citizens of Georgetown had three churches: the Baptist, the Methodist and the Episcopal.[139]

Economically, Reconstruction called for diversification of the economy in the form of introducing new businesses, but the era of the "New South" (the period after Reconstruction) is the era that witnessed the stabilization of those great economic changes. Property in land was no longer the basis for power, and the old elite lost their grip on society as a new group of leaders emerged as proponents of "New South Industries." Among the new industries to take root in the South during this era were textiles, tobacco, coal, lumber, railroads and extractive industries. Although the Georgetown rice planters continued to cling to the past to search for answers to their current problems, the economic landscape was in the process of change.

In Georgetown, "New South" industries steadily expanded their influence and grew to command the economy. Railroad companies, public utilities, shipping (which led to renewed interests for river and harbor improvements), banks, lumber companies, brick production and rice mills became the new foundation of the economy in Georgetown. The rice industry had met challenges from rice production in other states during Reconstruction, but in the post-Reconstruction world, other industries challenged the dominance of the rice planters at home and steadily gained influence over them throughout the Lowcountry.

The rice planters' wealth and influence passed on to rice millers, who built mills on several of Georgetown's major rivers. P.R. Lachicotte and

Sons built the first rice mill at Waverly Plantation on the Waccamaw River. The company took a new and unique approach: incorporation. In 1879, the Georgetown Rice Milling Company, a joint-stock company, built its mill in the town itself. During the late 1890s, the Guendalos Company worked the Pee Dee plantations, and S.M. Ward Company milled the Santee River crop.[440] No longer did rice planters themselves pound the grain of their own fields and their neighbors' fields. Instead, large consolidated outfits milled the harvest.

The Georgetown Rice Milling Company, the largest of the lot, combined the efforts of the local merchants and planters to solve labor problems. Built in town on Wood Street, it faced the Sampit River for easy access to the port.[441] In 1879, the company earned immediate respect when it received the highest price awarded for its sample of rice at the Charleston market. The following year, the company distributed a pamphlet to planters advertising its method of milling, which resulted in "less breakage of the rice in pounding."[442]

Other than labor problems, development and the competition from the new mechanized tools, a series of environmental catastrophes sped along the decline of Georgetown's rice culture. In 1893, two destructive hurricanes hit the region. Violent storms continued the following year and again in 1898, 1906, 1910 and 1911. During the spring thaws of 1906, 1907 and 1908, severe freshets destroyed crops and took their toll on the land by washing away rice banks.[443] Between 1893 and 1913, Georgetown's rice culture completely collapsed.

Among these scenes of economic despair, the people of Georgetown and the American South in general entered into an era of glorifying the "Lost Cause." Of course, veterans had been meeting in South Carolina to recount the war since its end, and some early monuments were erected, but by the 1890s, celebrating the "Lost Cause" had become a movement with followers proudly displaying the flags of the Confederacy and preachers leading their congregations in glorifying soldiers, generals and their roles in the war and even writing about their chivalrous efforts.[444]

By 1880, thirty-four-year-old Arthur Middleton Parker and his wife, Emma, were living at Mauricena Plantation on the Sampit River, which they had purchased from the estate of his maternal grandfather, Reverend Maurice Harvey Lance, who had died in 1870. After Lance died, his second wife, Sarah Laura, moved away from the area.

Just seven years after the death of Reverend Lance, Ester Jane Lance Read, wife of John Harelston Read II, died. She had been widowed in

1866 after the death of her husband. In 1871, Ester Jane married Ralph M. Izard in a private wedding. She died at the age of fifty-four and was buried on March 26, 1877.[145]

Arthur Middleton Parker was still listed as a rice planter in the 1880 population census. His thirty-two-year-old wife, Emma, was listed as a housekeeper, and his two eldest children, Mary and Frank (nine and seven years old, respectively), were listed as students "at school." Baby Arthur Middleton was one year old and Izard one month. The family had four house servants at that time: Louisa Ashe (fifty-five-year-old female), Hypus Lucas (sixty-year-old male), Patsy Lucas (twenty-year-old female) and Allen Ladson (thirty-two-year-old male).[146]

The 1880 agricultural survey for Arthur Middleton Parker at Mauricena Plantation shows the agricultural pursuits and status of that estate. The plantation consisted of 125 improved acres and 900 acres of woodland and forest with a farm value—including all buildings, fences and land—at $6,000. The value of Parker's farm implements was $20 and the value of his livestock, which included sixteen working oxen, $400. The amount paid in wages for the year was $242, and the estimated value of all products sold or consumed was $4,000. The main agricultural yields were eighty thousand pounds of rice and twenty bushels of Indian corn.[147]

Back on the Black River, Mansfield Plantation struggled to turn a profit. Thirty-year-old James Rose was the head of the household. On February 17, 1881, James Rose married Penelope Bentley Pyatt of Rosemont Plantation on the Black River at Mansfield Plantation. After the wedding, James and Penelope moved to Waterford Plantation on the Black River and began to build their own Black River rice empire by purchasing Deer Hall and Greenwich Plantations. On January 15, 1882, James and Penelope baptized their only son, James Rose. Together, the extended Parker family attempted to maintain their tradition of planting rice.[148]

Rutledge; his aging mother, Mary (now sixty-two and listed as housekeeper); and his family lived at Mansfield. Rutledge, listed as a twenty-nine-year-old planter; his twenty-nine-year-old wife and housekeeper, Charlotte Meade; their two sons, Rutledge (three) and John Emilius (born October 6, 1879); their daughter Anna Huger (born March 26); and their adopted daughter Agnes H. Burgess (nine) continued to plant rice. In 1882, the couple had their last child, Mariamne Meade, on June 30.[149]

The 1880 agricultural survey shows that the family cultivated 525 acres of the 825-acre estate. Mansfield Plantation was valued at $25,000 at that time, and the livestock value and the value of machinery were $300 and $200, respectively. The Parker family estimated their income at $12,000 for the year 1879, from which they paid an estimated $10,000 in wages and another $1,000 in plantation maintenance. After subtracting land taxes, the family made less than $1,000.

As the staple crop of the plantation, Mansfield yielded 450,000 pounds of rice. Supporting crops consisted of fifty bushels of Indian corn produced on five acres and fifty bushels of potatoes on a half acre. Livestock on the estate at the time of the survey consisted of three horses, seven mules, ten working oxen, two milch cows and six swine.[150]

During the 1880s, the family remained devoted to Prince George Winyah Church. Following family tradition, Arthur Middleton served as a vestryman from 1873 to 1884, and younger brother Rutledge served as a vestryman from 1878 through 1880 and then as a warden from 1880 until his death.

The 1880s were times of great loss for the Parker family, still grief stricken from losing Julia Rose in 1879. Mary Lance Parker died in March 1880, and Daniel Hayes died on March 24, 1881. The family buried Daniel the following day. He was twenty-five years old. Next, James Rose died at Mansfield on May 29, 1884. Two years and five months later, on September 6, 1886, Rutledge's wife, Charlotte Meade Parker, died as well. In 1885, the family's matriarch and last surviving immediate member of the Lance family, Mary Taylor Lance Parker, died at Mansfield. Lastly, Rutledge passed away on March 31, 1888, at Mansfield. He was buried the following day.[151]

With the passing of Rutledge, Arthur Middleton Parker became the sole proprietor of Mansfield Plantation. He and his wife, Emma, moved home to Mansfield, where he raised his children (Mary Taylor, Frank, Arthur Middleton and Ralph Izard "Kit") and those of his deceased brother, Rutledge, and sister-in-law, Charlotte Meade Parker (Rutledge, John Emilius, Anna Huger and Mariamne Meade) and continued to try to turn a profit in rice. After a string of deaths, the sole surviving male members of the Dr. Francis S. and Mary Lance Parker family were Arthur Middleton and Frank (who lived in Alabama).[152]

By the 1880s, the age of rice production was quickly screeching to a close in Georgetown. Estimated value of all farm products in the county had dropped from $1,274,157 in 1870 to $530,283 in 1880. The number of farms the size of Mansfield was quickly dropping in as well. In 1880, only sixty-nine farms in the county were between 500 and 999 acres. The value of all farming implements and machinery did increase from $15,862 in 1870 to $51,666 in 1880, but this was just an increase in the cost of doing business as planters invested their meager profits in an attempt to revive their devastated rice culture.[153]

Another challenge to holding on to the family estate came in 1883 with the formation of the Georgetown Land Association. The association was formed as a joint-stock company with $15,000 in capital. The organization's first project was to build a subdivision on the Serenity Plantation on the Sampit River. Soon, the city expanded to make development projects out of many Sampit and Black River plantations. Three years later, this same group started the Georgetown Building and Loan Association.[154]

In this economically overwhelmed state, new competition to the Georgetown rice culture appeared. Far to the southwest lay competition that, during this period, well surpassed the production of Georgetown and South Carolina on the whole. Louisiana was first to issue a challenge to South Carolina's rice dominance and quickly surpassed the state's production. In 1881, when the railroad connected New Orleans to Houston, land speculators and midwestern farmers moved to Louisiana and used the methods of wheat agriculture to produce rice.[155] Western rice growers employed harvesting machinery, improved Japanese seed, steam-powered engines, gang plows, seeders, discs and twine binders to achieve dominance.[156] The Agricultural Revolution had finally come to rice production; however, because of its soil, Georgetown planters could not follow the lead of mechanized farming.

Louisiana's rise to prominence was achieved by the result of several factors. The main reasons for the increased production of rice in Louisiana were due to the collapse of the state's sugarcane industry and, as aforementioned, the rise of the railroad.[157] Rice production in Louisiana was less labor intensive than sugar production and, consequently, attracted more laborers. Another factor was the drop in land value. Land had plummeted by more than 70 percent since the Civil War, thus enabling many farmers to purchase land. The state experienced a serious decline in the production of sugarcane after the war and met serious competition from Cuba and South America, which quickly overwhelmed Louisiana growers. With their number-one cash crop collapsing around them, Louisiana farmers found rice a viable solution.[158]

By the late 1880s, Texas and Arkansas, stimulated by the Southern Pacific Railroad and aided by machinery, developed the "Prairie Rice Culture." By employing advancements made in Louisiana, Arkansas and Texas boasted that for each tractor that they employed, they could break fifty acres of land per day.[159] Although production in these states did not offer a threat to the consistent Georgetown rice crop of the 1880s, by the early 1900s, their production had dwarfed that of South Carolina, North Carolina, Georgia and Louisiana combined. In 1903, Texas, Louisiana and Arkansas produced 99 percent of America's rice crop.[160] By 1909, Texas had irrigated 286,847 acres of rice, and South Carolina's contribution to the country's exportation of the crop ceased to bear significance.[161]

From the 1880s through the early 1900s, everything about the rice industry declined in Georgetown. The number of farms in Georgetown decreased from sixty-nine in 1880 to sixty in 1890 to forty-one by 1910. The value of all farm products never truly recovered after the war, and by 1900, the estimated value of all farm products produced in Georgetown was only $532,626, less than half the 1870 value of $1,274,157, which in itself was only a small fraction of the antebellum greatness. The value of all farming implements and machinery had dropped to $80,580 by 1900, about one-eighth of the prewar value of $616,774.[162]

With the profitability of the rice culture collapsing around them, and in an attempt to consolidate their economic interests or perhaps keep Mansfield out of foreclosure, Arthur Middleton and Emma Izard Parker sold Mauricena Plantation on the Sampit River in 1894. It was soon thereafter subdivided into a housing development.[163]

It was during this era that the Sons of the Confederate Veterans and the Daughters of the Confederate Veterans organizations formed. Also, citizens throughout the South continued to build monuments to glorify their fallen fathers, brothers, uncles and kinsmen from the South's failed independence movement. Each local county built a monument, and Georgetown's was dedicated to the Georgetown Rifle Guards, Company A of the Tenth South Carolina Regiment, in 1891. The monuments were originally erected in cemeteries but were later moved to central locations in towns, where they would be more visible.[164]

Glorification of the Confederate cause and dead were honorable notions, but these actions of gratitude did not pay the bills. With the rice culture collapsing around them, many in the area committed the county to "New South" industries and, for the most part, gave up on farming. These new extractive industries such as lumber, turpentine and phosphate ravaged the land.

The Atlantic Coast Lumber Company became the most profitable of these extractive industries. Originally founded in 1899 with a capital base of $1 million, the company had built a complex, including four sawmills and all the necessary appendages, by 1903.[165] Capable of sawing 300,000 feet of wood per day, it attempted to buy or at least rent many of the plantations. By 1905, the company was exporting 300,845 tons of lumber for a value of $2,645,537, more than 25 percent of all exports from Georgetown.[166]

By 1913, the company owned or held rights to 250,000 acres of land in eight of South Carolina's counties. To log its lands more efficiently, the company built its own railroad, the Georgetown & Western Railroad. The rail ultimately had 317 miles of track, which it used to bring its lumber to the mill.[167] At its height, the company employed eight logging crews, who brought their housing with them as they traveled from location to location.

At the turn of the century, rice continued to lose influence throughout the Lowcountry as South Carolinians planted 35,041 acres in 1901. This land-loving but ever-shrinking group resented and often complained about the mills, the loggers and others involved in extractive industries that stripped the land of its resources.[168] Unfortunately for the environment, some Georgetonians refused to listen to the arguments, mistakenly interpreting the immediate cash return of the extractive industries as progress.

Elizabeth Allston Pringle, the owner of Chicora Wood and White House Plantations on the Pee Dee River, frequently complained about the Atlantic Coast Lumber Company's actions in the column she wrote for the *New York Sun*. She mentioned being propositioned by loggers, both through letters and conversations, to sell logging rights to her land. Although Pringle was nearly bankrupt, she never allowed the loggers to cut on her land. In 1903, she wrote:

> *I got a note from Mr. L*[achicotte] *asking me to allow four negro men to cut 100 cords of wood on my land and he would be responsible for the money, $25…I heard afterwards that a neighbor had sold them the right to cut on their land, and when I went to the landing to-day I saw about fifty cords of the wood they had cut and piled there, and it was the most splendid fat lightwood I ever saw, from trees that had been growing*

> *on that land sixty or seventy years. And the owner gets 25 cents a cord, while the wood brings $2 anywhere.*[169]

When the planters refused to sell their timber, the company simply propositioned the tenants, who found it hard to refuse the money. Pringle recalled catching two of her "trusted" tenants chopping wood for personal gains in 1906.[170]

Pringle was one of the last rice planters to produce rice for export in Georgetown County. Between her two plantations, she planted a combined total of 200 acres of rice in 1903. Of her planted acres, 50 were wage paying, while the remaining 150 were farmed by her tenants. Pringle decreased her planted acres to 150 in 1904, and by 1906, after her crop was destroyed by freshets, she quit planting rice and diversified her plantation.[171] Like many other planters in Georgetown, she began planting cotton, rye, peas, oats, strawberries, potatoes, corn, alfalfa and other produce.

Like most of the plantations in the county, Mansfield Plantation fell victim to the extractive industries as Arthur Middleton Parker desperately tried to hold on to his beloved family plantation. As a last resort, timber and turpentine were taken from the highlands, but still Parker could not make farming the old estate profitable.[172] In 1906, news arrived that Arthur Middleton's last surviving brother, Frank Parker, had died in Mobile, Alabama, and was buried there in Magnolia Cemetery. The following census report (1910) showed that Arthur Middleton (age sixty-five) and his wife, Emma (sixty-three), lived alone in the Mansfield manor house. Arthur Middleton Parker was no longer considered a "planter"; instead, the designation "farmer" was added after his name to distinguish his chosen profession.[173]

In 1912, when most of Georgetown's rice lands lay fallow, when all hope of planting was gone, after the timber and turpentine had been extracted from the forest and the family had nothing left to sell off of the old farm, Arthur Middleton Parker passed away while vacationing in Flat Rock, North Carolina.[174] Left without the direction and guidance of the patriarch, the surviving members of the family decided to sell Mansfield. The heirs of the Rutledge Parker family (Rutledge Parker of Montana; Dr. John E. Parker of East Orange, New Jersey; Anna H.P. Morris of Cumberland, Pennsylvania; and Mariamne Mead Parker of Charleston, South Carolina), the heirs of Frank Parker (Eliza Troost Parker of Mobile, Alabama, and Frank Troost Parker) and Arthur Middleton's

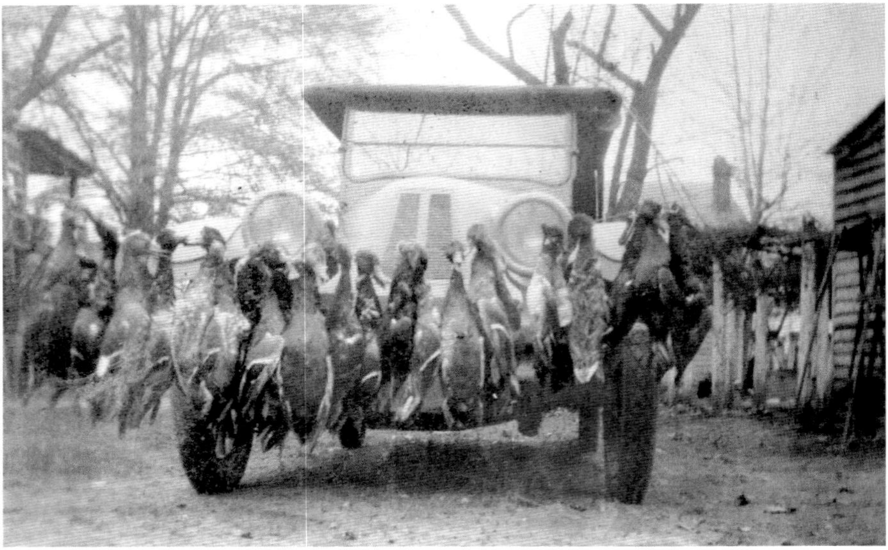

The rice fields once teemed with wild duck. This string of ducks is the result of a hunt at Mansfield.

widow, Emma Izard Parker, hired E.N. Beaty and N.L. Smith to survey the estate. The surveyors drew up a plat showing twelve former slave cabins, the manor house and five outbuildings surrounding the manor house. Later that year, on July 15, 1912, they sold the 760-acre estate (250 acres of rice land and 510 acres of pine land and highland) to Charles W. Tuttle of Auburn, New York, for $8,500.[175]

No longer having ties to the area, the heirs of Frank Parker and Rutledge Parker went to their respective homes, while Emma moved to Flat Rock, North Carolina, then to Ardmore, Pennsylvania. She died in a car crash on May 21, 1937, and is buried at Saint John in the Wilderness Cemetery in Flat Rock.[176] Charles W. Tuttle took possession of the old plantation, turned the property into a hunting club and used the manor house as a winter residence.[177]

Also in 1912, John Harleston Read III died at Maryville Plantation. Just like their cousins the Parkers, the Reads decided to sell the seat of their personal rice empire out of the family's ownership, signaling an end to the extended family's rice interests.

By 1913, the Georgetown rice culture had ceased to be profitable. Some farmers continued to plant rice for personal consumption and local sales but not for national export. The name "Carolina Rice"[178] continued to be placed on the staple crop; however, the center of crop production shifted southwest to Arkansas, Texas, Louisiana and, later, California.[179]

The following spring, fields lay fallow. Damaged floodgates and banks were ignored, and farm tools began to rust. Docks and rice barges stood vacant, and winnowing houses and threshing and pounding mills remained locked. The work songs of the laborers, once heard throughout the countryside, ceased, and there was nothing to break the stillness of the boggy Lowcountry.

Over time, the rivers and swamps began to reclaim the rice kingdom. Cypress trees sprouted out of the once nurtured fields, and alligators and snakes returned to the abandoned lands. In the spring, the ricebirds made their annual visit but paused for only a short time before flying north. The empire built on rice had fallen. After more than 150 years of family occupancy, Mansfield had fallen out of the Man-Taylor-Lance-Parker family hands seemingly forever. Defeat in their bid for independence had given rise to a new class of wealth in the region: the industrialist and their banker friends who had eclipsed the great wealth of the founding families of the American South.

*Epilogue*

# The Next One Hundred Years

*History is a symphony of echoes heard and unheard. It is a poem with events as verses.*
*—Charles Angoff*

For two decades, Charles W. Tuttle owned Mansfield and used it as his own personal hunting resort. In the early 1900s, Mansfield's woods were filled with wild game and duck.

In February 1932, the Tuttle family sold Mansfield Plantation to Colonel Robert L. and Charlotte Hope Binney Tyler Montgomery of Androssan, Pennsylvania, for $30,000.[480] Shortly after their purchase, the Montgomerys began restoring and modernizing the plantation. The colonel's employees created imposing brick entrance gates, as well as matching brick bridges and walls along the entry road; bricked up the surviving outbuildings; and created brick walkways using recycled bricks from the adjacent plantations. Mansfield now holds pieces of the surrounding antebellum community and legends of the surrounding area plantations.

The Montgomerys added the south wing to the manor house, as well as the basement, which includes a dumbwaiter and small bar for entertaining. They added basements and brick veneer to the kitchen and schoolhouse and converted them into guest quarters. Concomitant with these activities, they also constructed the north guesthouse, which matches the design of the restored structures. They replaced the wooden shingles on the antebellum slave cabins; added two additional structures

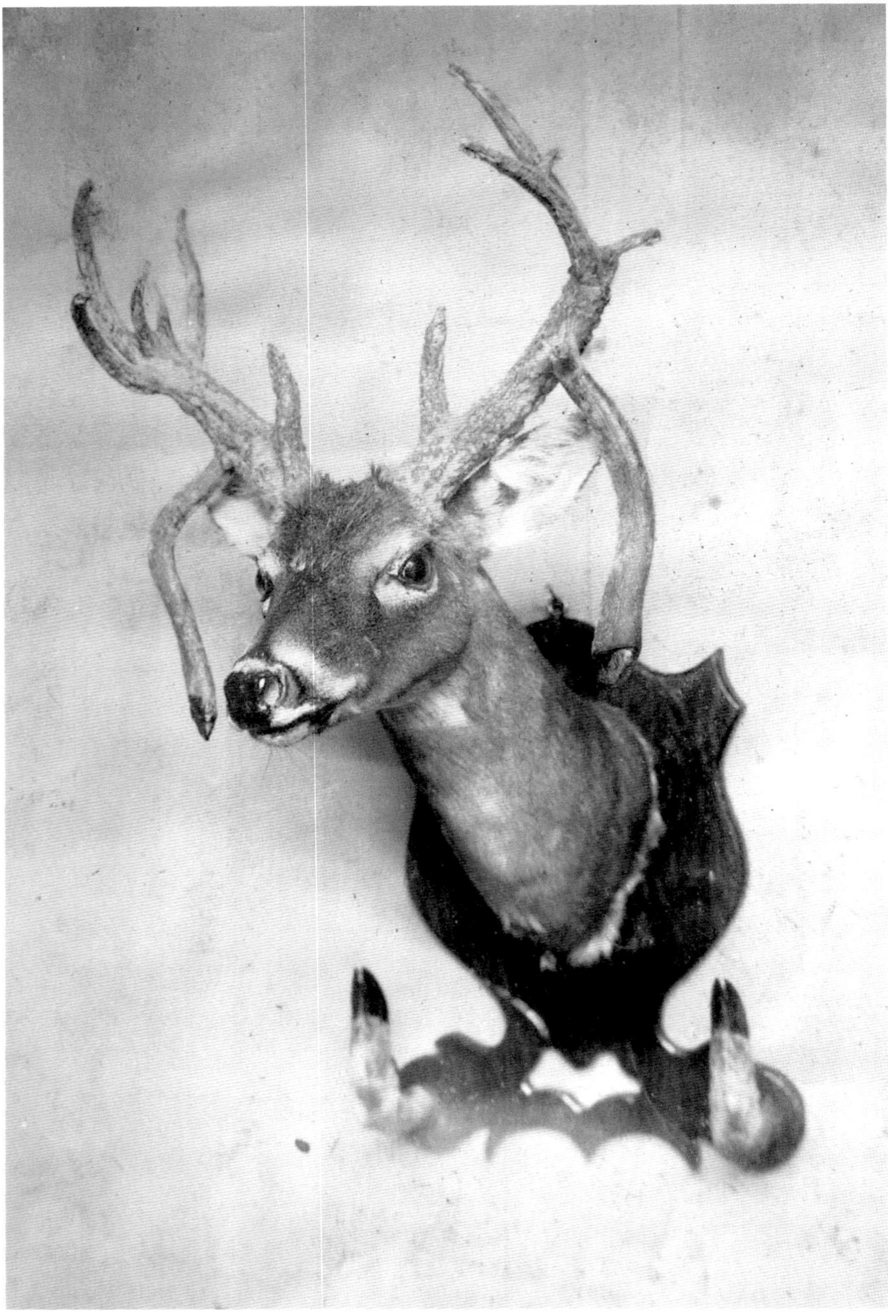

This large buck was hunted at Mansfield.

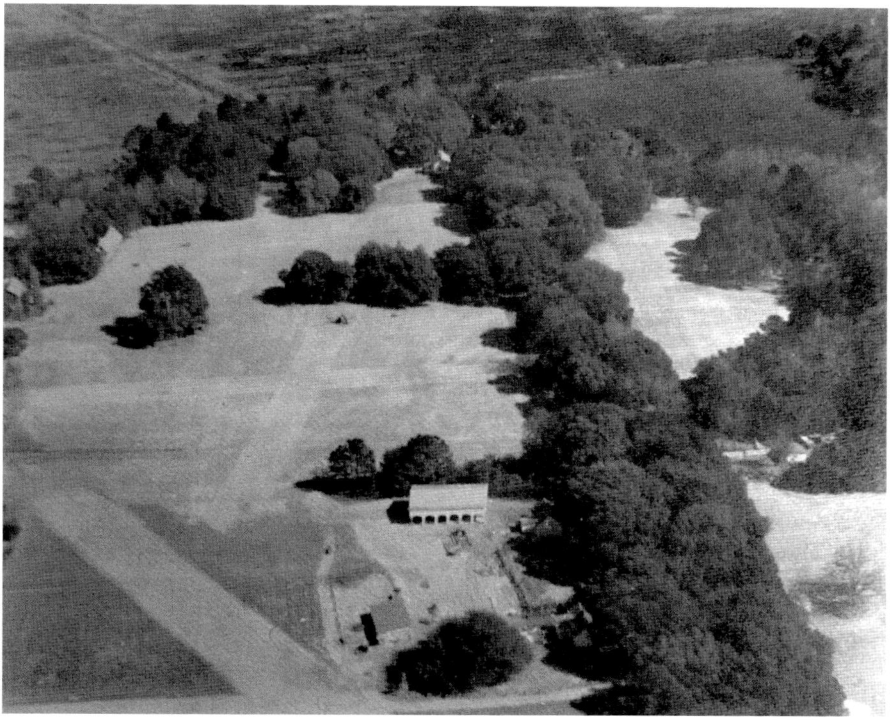

This aerial view of the estate clearly shows Colonel Montgomery's landing strip.

to the village, including a laundry; and constructed a horse stable, a landing strip and a hangar for their personal airplane.

The Montgomery family not only restored the look of the plantation to their satisfaction but also brought some of Franklin Roosevelt's New Deal to Georgetown by hiring more hands and revived the production of rice. Beginning in 1939, Colonel Montgomery produced a model rice crop that his family, their friends and employees consumed.

In 1943, due to wartime shortage of labor, Mansfield cultivated its last rice crop. At this time, the Montgomerys constructed a lake (which became a haven for wild game) by impounding portions of the rice fields located behind the main house.[181]

Colonel Robert L. Montgomery died at Mansfield in January 1949, and Mrs. Montgomery passed away on June 6, 1970. Without their collective stewardship, the 1840s settlement pattern (constructed under Dr. Francis S. and Mary Lance Parker) would have been lost to the elements as so many antebellum plantations were during this era. On

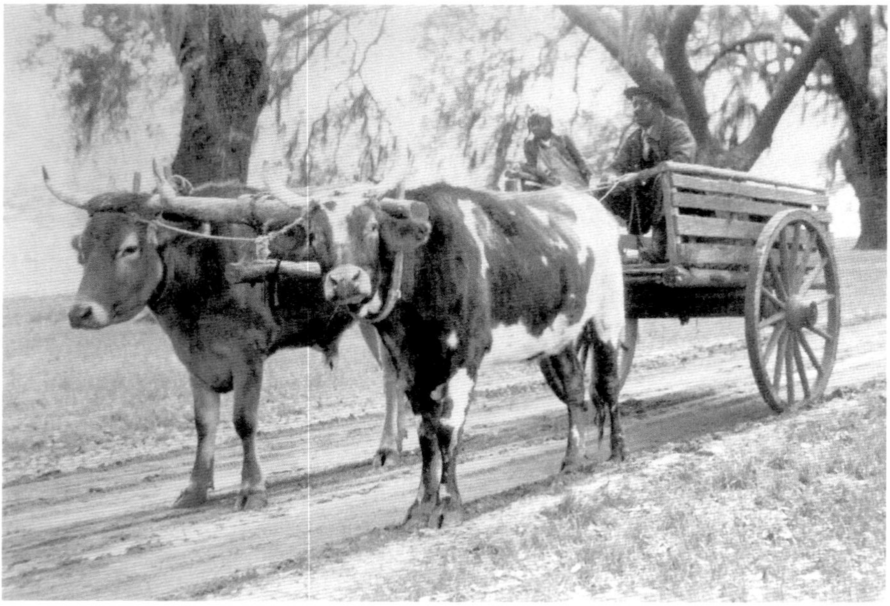

Anonymous village dwellers driving an ox cart down Mansfield's street in the late 1930s or early 1940s. The child is wearing aviator goggles. Perhaps the goggles were a gift from Colonel Montgomery.

October 28, 1970, R. Alexander Montgomery and Robert Montgomery Scott, the executors of Mrs. Montgomery's will, which she wrote on May 23, 1967, sold the 760-acre plantation to Wilbur Stevenson Smith of Columbia, South Carolina, for $405,500.[182]

Smith, who used Mansfield as a summer residence, grew soybeans, corn, beans and rye on the property.[183] In 1977, Smith began the process of adding Mansfield Plantation to the National Register of Historic Places. He further protected the legacy of Mansfield Plantation by donating Dr. Francis S. Parker's plantation record books to the South Caroliniana Library at the University of South Carolina, where they could be preserved and protected until properly analyzed.[184] Smith died in 1990, at which time his estate was conveyed to his daughter, Sarah (Sally).[185]

Sally Smith owned Mansfield for a dozen years and brought much publicity to the plantation and its glorious history by opening the estate as a bed-and-breakfast event venue. To further the bed-and-breakfast, she added a new kitchen and an addition on the backside of the north wing.

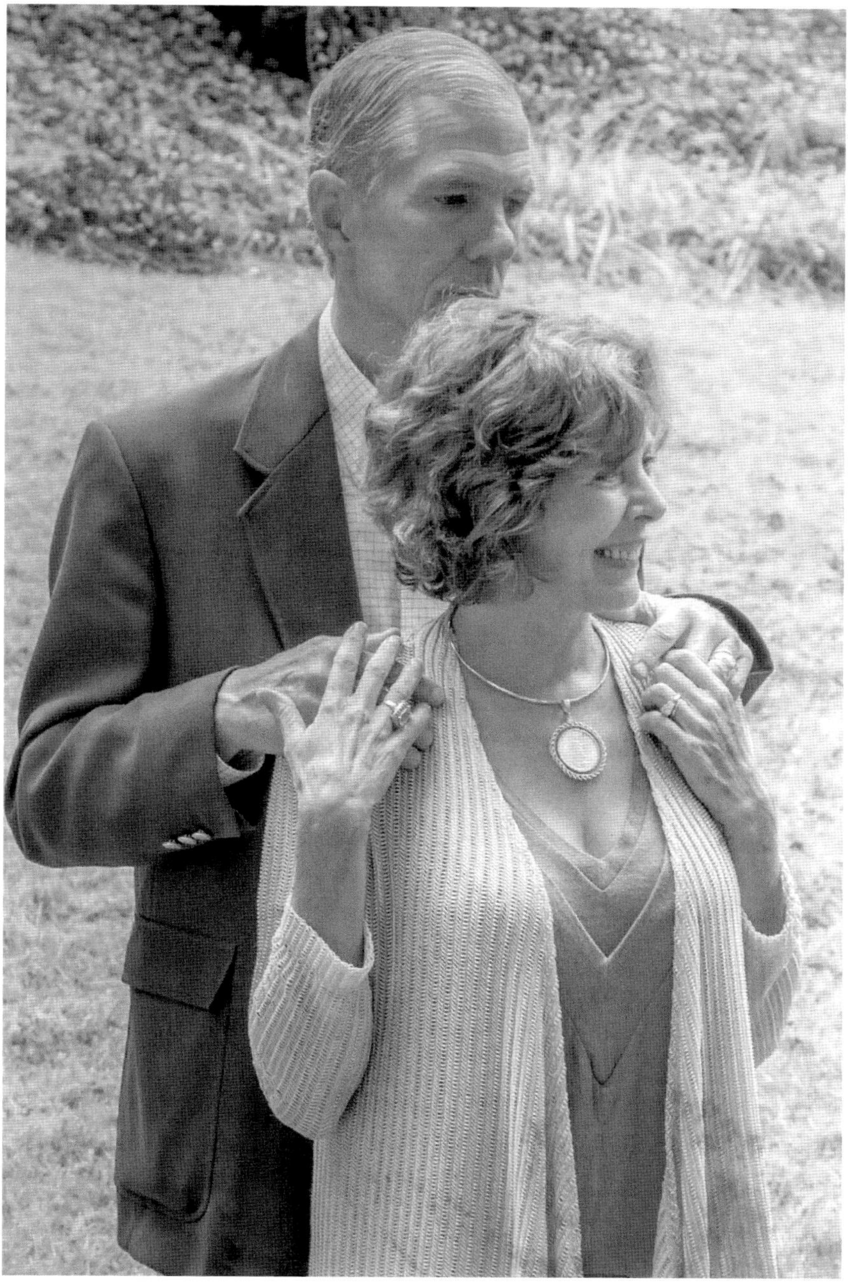

John and Sallie Middleton Parker are the current owners of Mansfield Plantation. Under their stewardship, the plantation is being restored and preserved for future generations to enjoy and study.

Sally, a PhD historian and house museum director before she moved to Mansfield, showed a deep interest in historic preservation and historical analysis. She funded the first archaeological survey of the estate, which was conducted under James L. Michie of Coastal Carolina University. Sally's goal, beyond historical interpretation and the development of a bed-and-breakfast resort, was the continual preservation and stabilization of the existing plantation and its structures.

In 2005, Sally Smith sold Mansfield to John and Sallie Middleton Parker, thus returning the estate to the Man-Taylor-Lance-Parker family. The Parkers continue the model of stewardship to Mansfield Plantation that the Montgomery family began in the middle of the twentieth century. The Parker family's connection to the land is great, and their commitment to preservation and conservation can be seen through their work and the work of their board of directors. Sharing a perpetual labor of love and understanding, the Parkers and their board work to reveal Mansfield's secrets, share our plantation past and honor all those who lived, dreamed, labored, loved, won and lost at Mansfield—planter and slave, farmer and employee, caretaker and guest.

*Appendix*

# The Architecture of Mansfield Plantation

*History is life; he who has not lived, or has lived only enough to write a doctoral dissertation, is too inexperienced with life to write good history.*
*—Louis Gottschalk*

## The Manor House

The central part of the house is the oldest and measures about thirty-eight feet square. It consists of four rooms. It is uncertain who built it, but consideration must be given to either Susannah Man or John Man Taylor. Although hand-wrought nails were being replaced with early machine-cut nails in 1803, when John Man Taylor returned from Harvard to be a rice planter, some plantation blacksmiths continued to use the forge and anvil to produce nails for the plantation's use. Therefore, either of these two early proprietors could have built the main section of the manor house.

The presence of late machine-cut nails, which became available in the 1830s, suggests that Dr. Francis S. Parker or Reverend Maurice Harvey Lance constructed the north wing of the house. In this section, the hardwood flooring matches the flooring in the main section, suggesting that they were floored at the same time. In 1941, an article in the *Georgetown Times* entitled the "History of Mansfield Plantation" claims

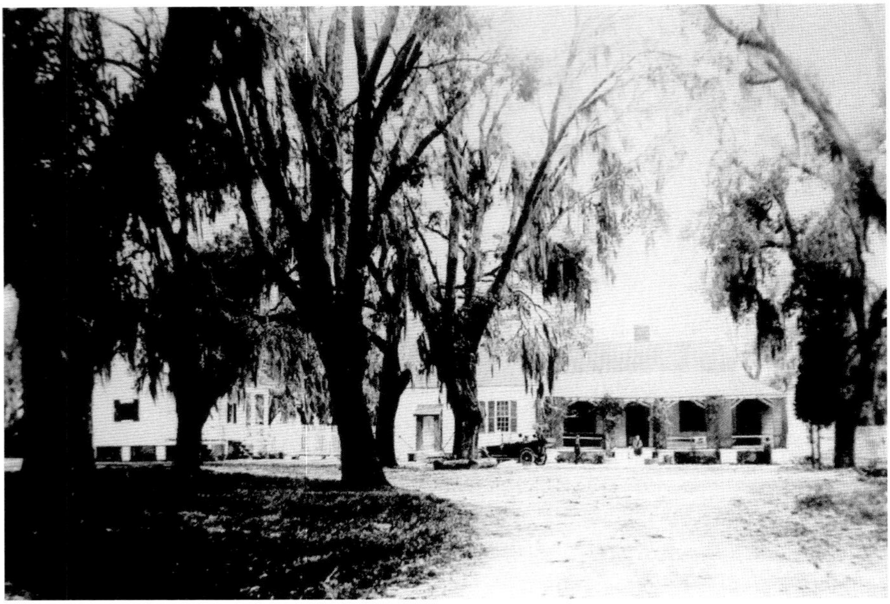

The manor house prior to Colonel Montgomery's addition of the south wing. This picture was probably taken during the Tuttle residency. The house had a different front porch and a tin roof and featured significant ivy growth. The schoolhouse is simply resting on blocks or pedestals, not on a full foundation as it is today.

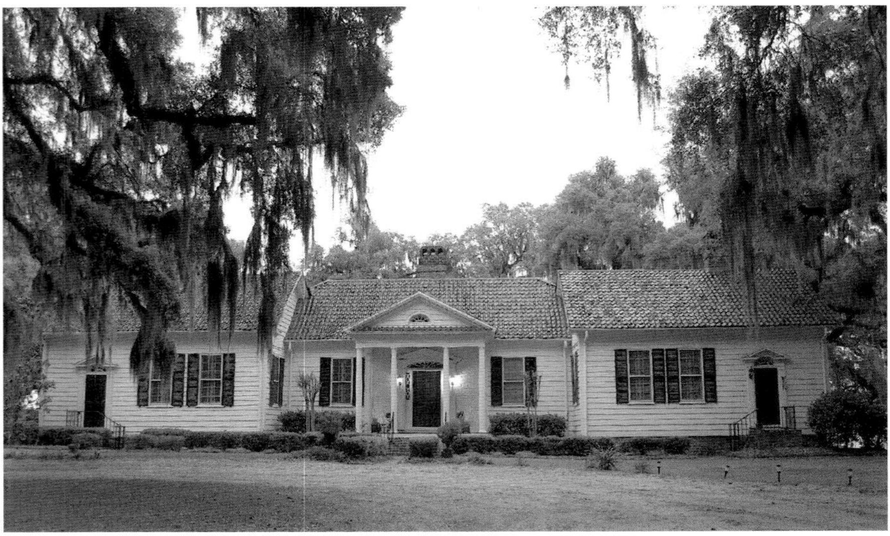

The manor house as it stands today.

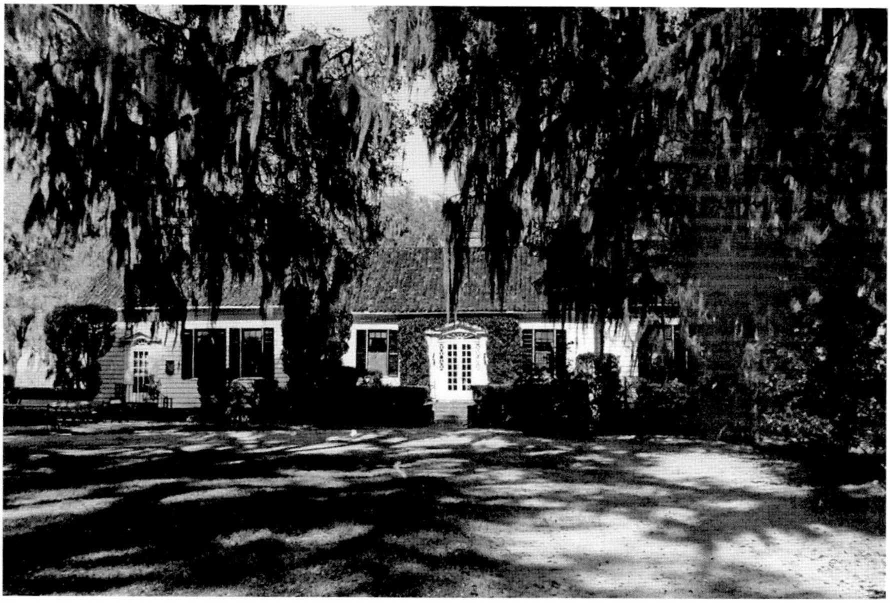

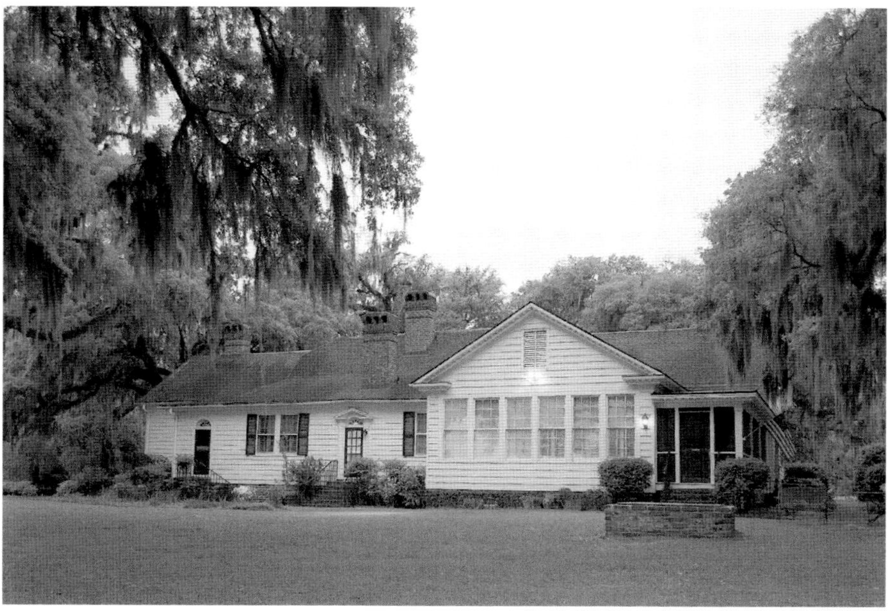

The back of the manor house today showing the Sally Smith Cahalan addition.

that Dr. Parker moved into the house and added the north wing. The rear part of the parlor (where the piano is now) has different flooring; it was probably an outside porch at the time of the Parker addition and closed in at a later date.

The Montgomerys modified the manor house to include a new addition to the south, a basement, a kitchen, several bathrooms and an altered chimney design. They lifted the house in order to add the basement, which accommodates a coal-fired furnace and a system of steam radiators. Plumbing was added for bathrooms and the kitchen, and refrigeration units, storage and laundry facilities and a small bar room were added for entertaining. They also altered the interior of the house with decorative details, including an altered fireplace and the addition of Georgian ceiling motifs and ornate door surrounds. Prior to the Montgomerys' extensive work, the house rested on brick piers with a singular addition to the north.

Exterior modifications by the Montgomerys included the replacement of the windows, window frames, sashes and trim; the application of clapboard; a ceramic tile roof on the front and tar shingles on the back; Georgian pediments and pillars around the entrance doors; a brick façade; and extensive brick walkways.

The north rear of the house received another addition under the ownership of Sally Smith.

## The Antebellum Kitchen

The antebellum kitchen originally had clapboard siding and glazed windows and rested on brick piers. Like the manor house, the 33.5- by 21.5-foot structure experienced modifications during the Montgomery ownership. The attic was converted into living quarters for some of the Montgomerys' servants, and they added a basement to the structure, which served as a storage place for coal, as well as plumbing and electrical systems. The Georgian pediments and pilasters surrounding the entrance were also added during the Montgomery ownership.

Colonel Montgomery added a full-service kitchen to the manor house and then converted the old kitchen into a three-room guesthouse with two rooms in the front of the building on the ground level and one on the

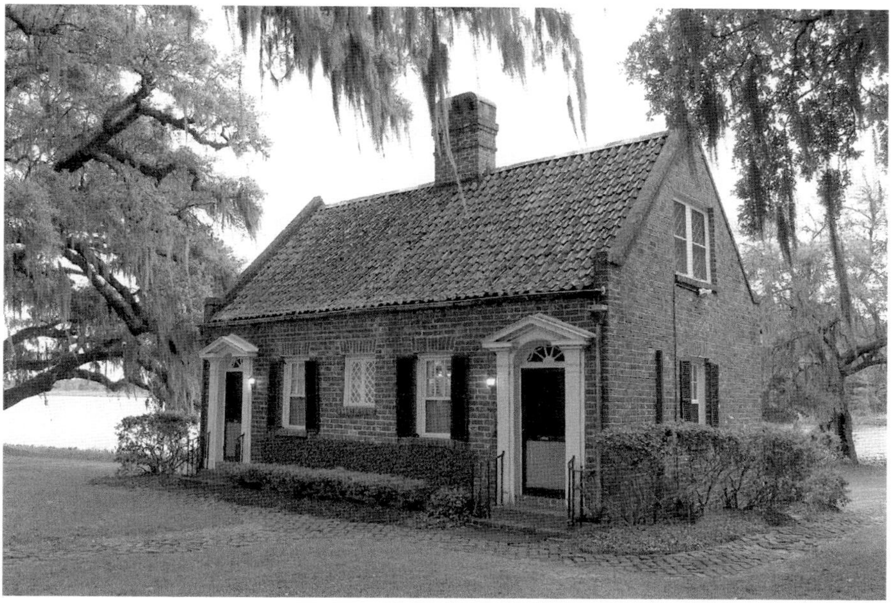

The old kitchen as it stands today.

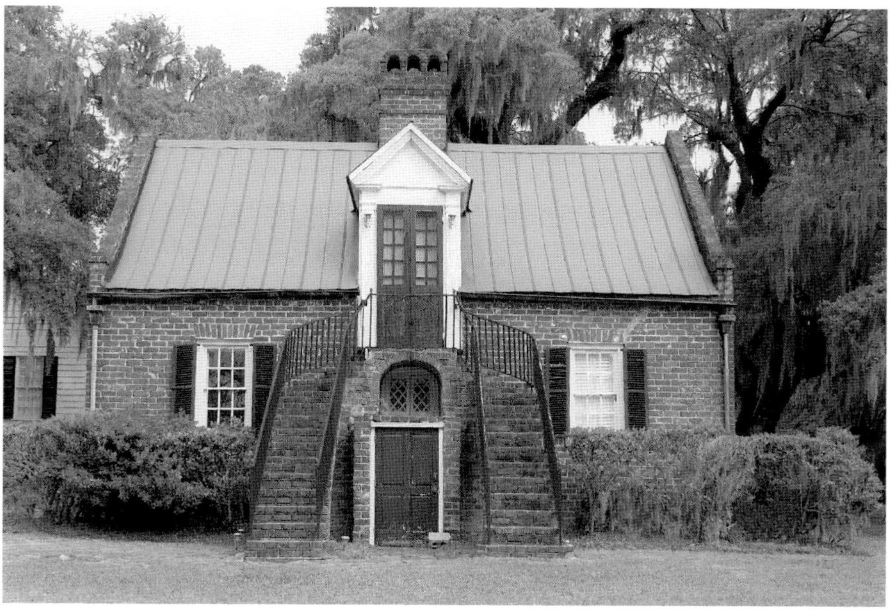

The back of the old kitchen.

second floor, which is accessed by a set of stairs in the rear of the structure. The east room on the ground floor has an original pencil signature of Dr. Francis S. Parker, as well as others, dated December 1843.

## The North Guesthouse

The Montgomerys built the guesthouse to accommodate the needs of their friends and family members who visited them on the estate. The guesthouse carries the same extensive brickwork and Georgian ornamentation found throughout the rest of the estate. Charleston contractor Thomas Pinckney built the ironwork staircase, which he designed to emulate the ironwork already in place on the old kitchen. The guesthouse contains two rooms on the ground floor, accessible in the front of the building, and two rooms in the rear of the building, reached by the ironwork staircase.

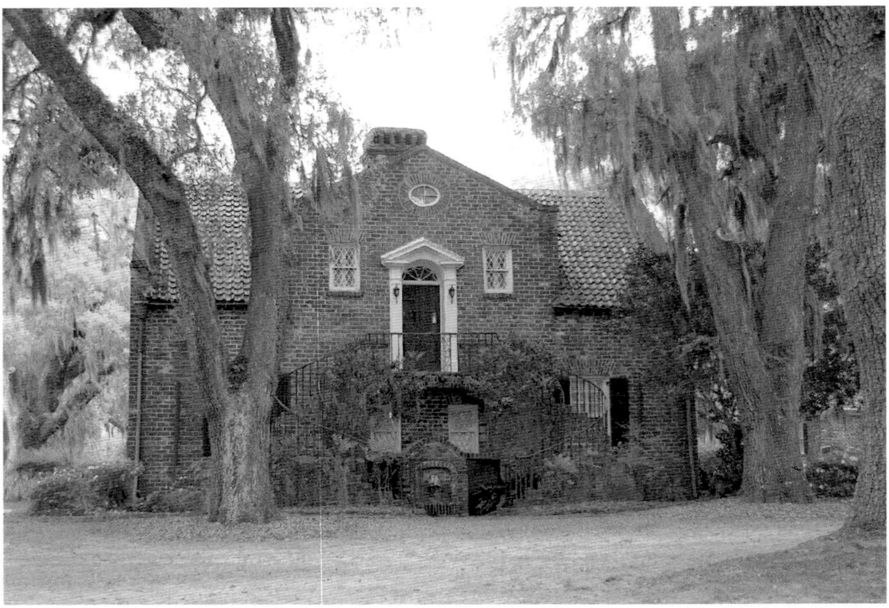

The north side of the north guesthouse.

# THE ANTEBELLUM SCHOOLHOUSE

Dr. Francis S. Parker built the schoolhouse for the education of his children. J.R. Parker stated in his *History of Mansfield Plantation* that the "school house or cottage" was built to accommodate the needs of Parker's expanding family. The original structure had a full-width porch that was originally built on brick piers, covered in clapboard and accented by glazed windows. The structure is 36.3 by 21.2 feet.

As with the kitchen, the Montgomerys added a basement in order to accommadate plumbing and electrical utilities. They also added nineteenth-century brick veneer and a typical Georgian cornice with dentils, which carries the theme prevalent in other structures on the estate. History does not record the name of any antebellum tutors, however; according to Parker family tradition, Arthur Middleton

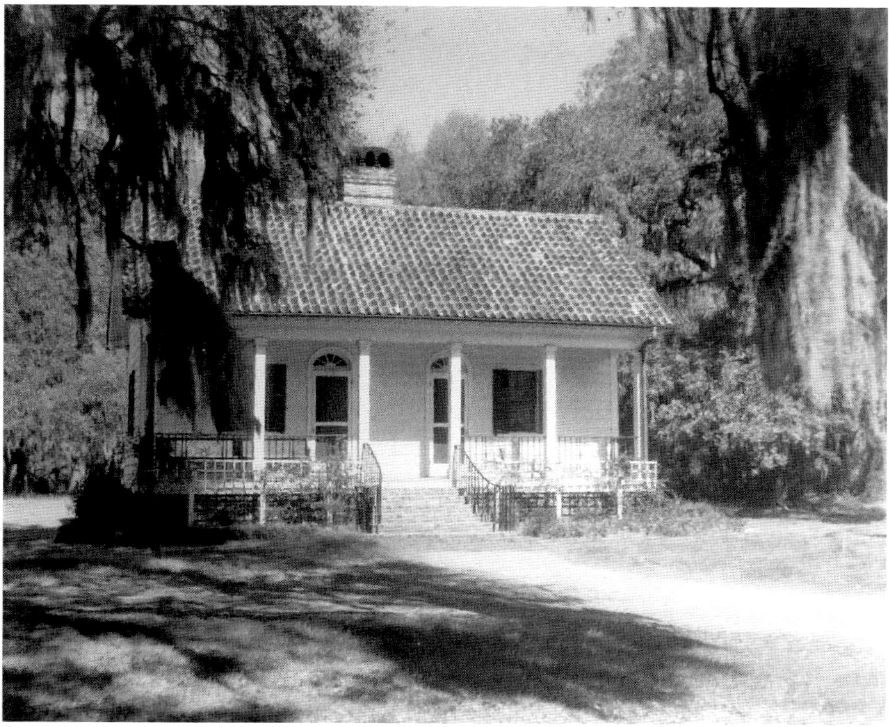

The schoolhouse prior to being bricked up during the Montgomery residency. It is, however, sitting on a full brick foundation.

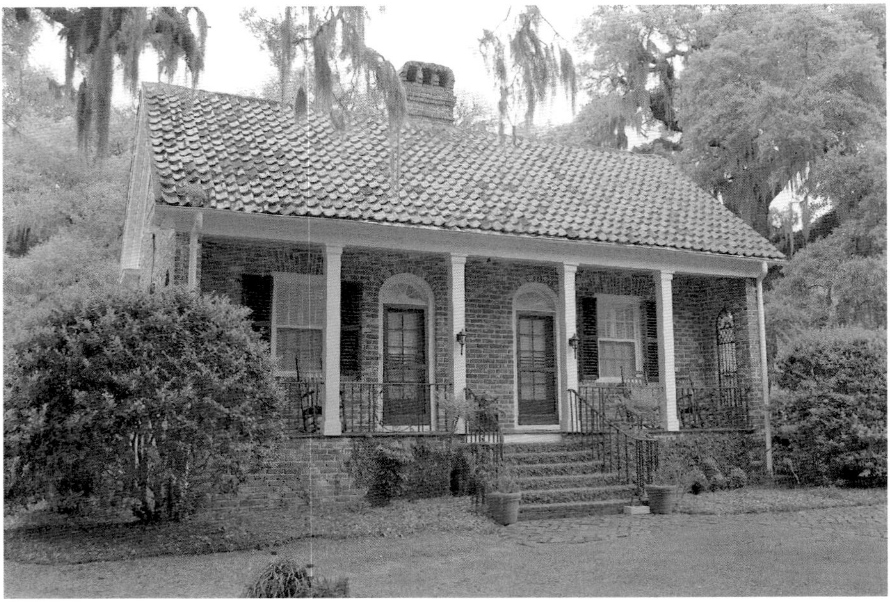

The schoolhouse as it stands today.

Parker employed a strict tutor named Mr. Valentine, who remained with the family for a number of years. Apparently, Mr. Valentine was a mathematical genius who pushed the Parker children in yearlong studies even while they summered in Flat Rock, North Carolina. Mr. Valentine personally believed that no education was complete without mastery of the quadratic equation.

## The Rice Mill

The rice mill might have been built under the ownership of Dr. Francis S. Parker or the Reverend Maurice Harvey Lance. Although threshing rice mills date back to the era of the Lance ownership, the value of the machinery and farm implements on the Parker estates increased dramatically between 1850 and 1860. This might have been the result of a mill built at each of his estates or the construction of a couple of these mills. The function of this mill was to separate the tailings and straw from the grain.

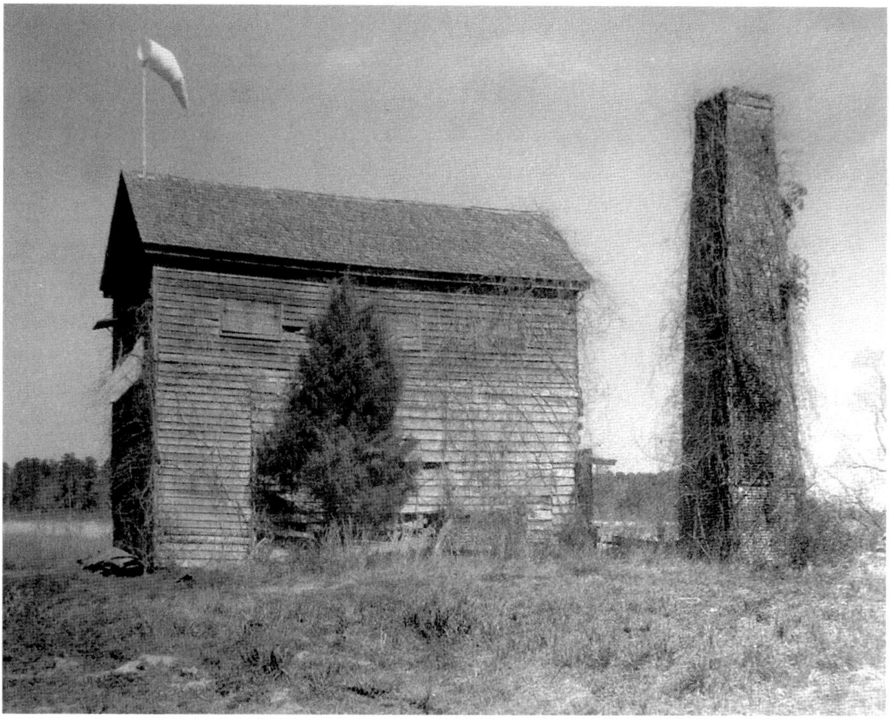

The original steam-powered rice-threshing mill during the Montgomery residency. Colonel Montgomery placed the windsock on the top of the mill to help him safely land his airplane at Mansfield during the 1930s and 1940s.

The mill, which accidentally burned in the 1980s, was a medium-sized, two-story, rectangular, wooden-framed building covered in clapboard siding. The chimney is eight feet square at the base and tapers gradually to about five feet. Its overall height is approximately forty feet.

## THE WINNOWING HOUSE

The winnowing house is the least-altered structure on the estate. The building was probably constructed during the Parker ownership in the 1840s. The siding is of clapboard, and the roof of the structure is covered in wooden shingles. The structure measures about twenty-three

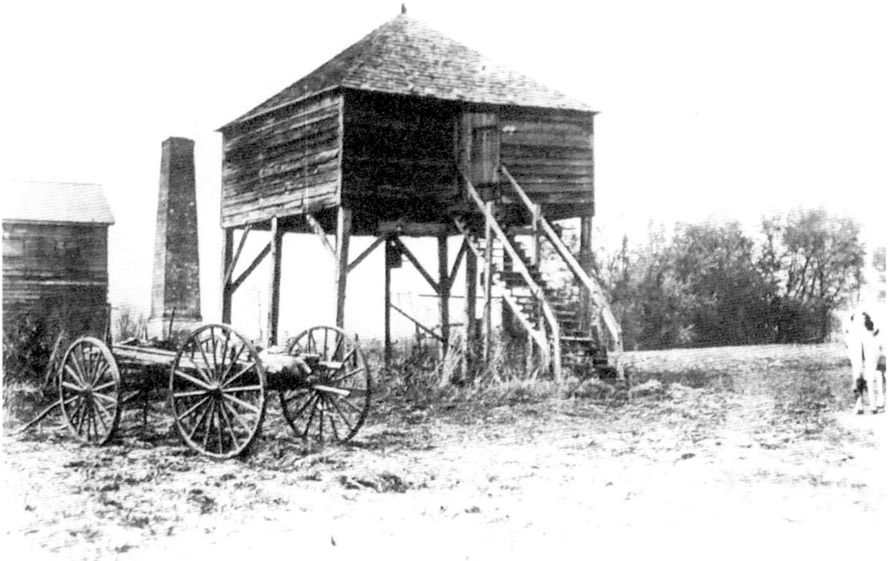

The staircase on the front of the winnowing house and the cart suggest that the structure was still in use for rice preparation at the time of the photo. This picture is from either the Tuttle or early Montgomery residency.

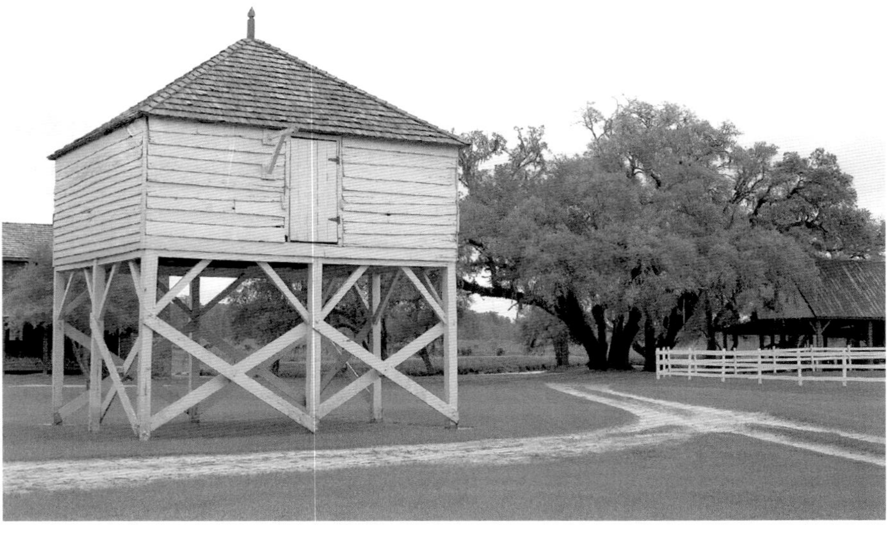

The winnowing house as it stands today.

feet square, and the floor is elevated fifteen feet above the ground on large wooden pilings.

The function of the structure was very simplistic. Laborers climbed an exterior set of stairs (since removed) into the winnowing house. Rice grain was hoisted up to the opening of the house with a boom and then swept into a hole in the floor when the wind was blowing. During the fifteen-foot drop from the floor of the winnowing house to the ground, tailings and chaff were blown off of the already threshed rice by the wind.

Support braces were added to the structure after hurricane Hugo to strengthen the building.

## The Slave Chapel

Like many other structures on the estate, the slave chapel was built under the ownership of Dr. Francis S. Parker. The slave chapel is a one-room building that is approximately the same size as the existing slave cabins on the street. It still retains some of the original clapboard siding, as well as the original wall studs, flooring and exposed roofing rafters.

The Montgomery family renovated the chapel to include a front porch, glazed windows, tar roof shingles and a small steeple with asbestos siding. They treated the interior with a raised pulpit, plastered interior walls and front and back doors. In 2005, John and Sallie Middleton Parker restored the chapel back to its antebellum simplicity and dedicated it to Rebecca Golden, a trusted family servant who grew up at Mansfield under the residency of Arthur Middleton and Emma Izard Parker. In 1912, after the death of Arthur Middleton, Rebecca moved with Emma Izard Parker and remained in service to the Parker family until her death. She helped raise John Parker, the current owner of Mansfield.

The chapel's bell was forged by the Gould Company of Seneca Falls, New York. The company produced only bells between the years 1887 and 1890. The thirty-three-inch bell might have been placed at Mansfield by Arthur Middleton Parker or perhaps bought as a used bell and added to the estate under the Montgomery renovation.

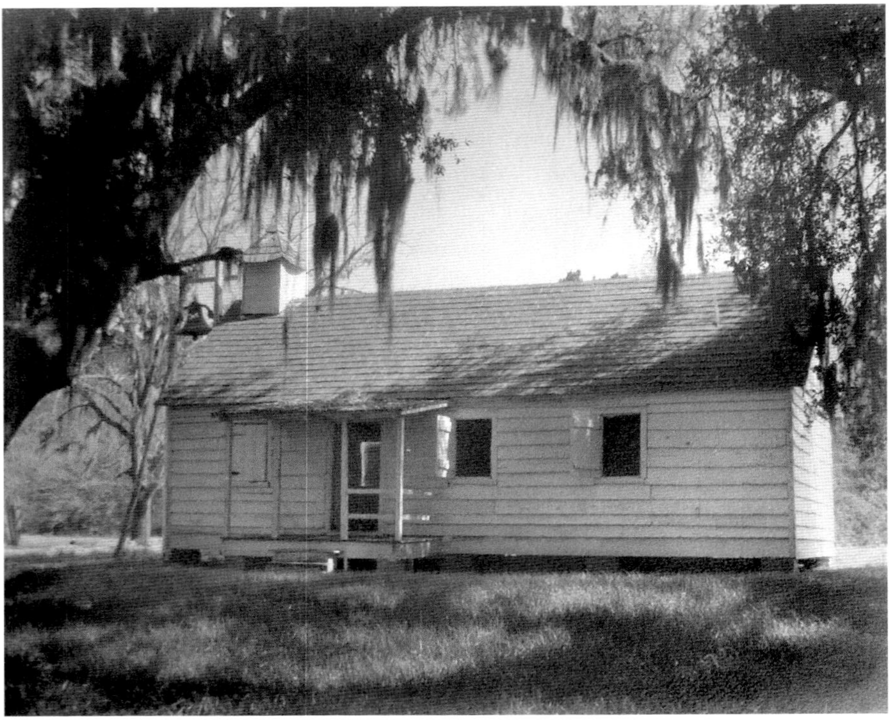

The chapel as it looked during the Montgomery occupancy.

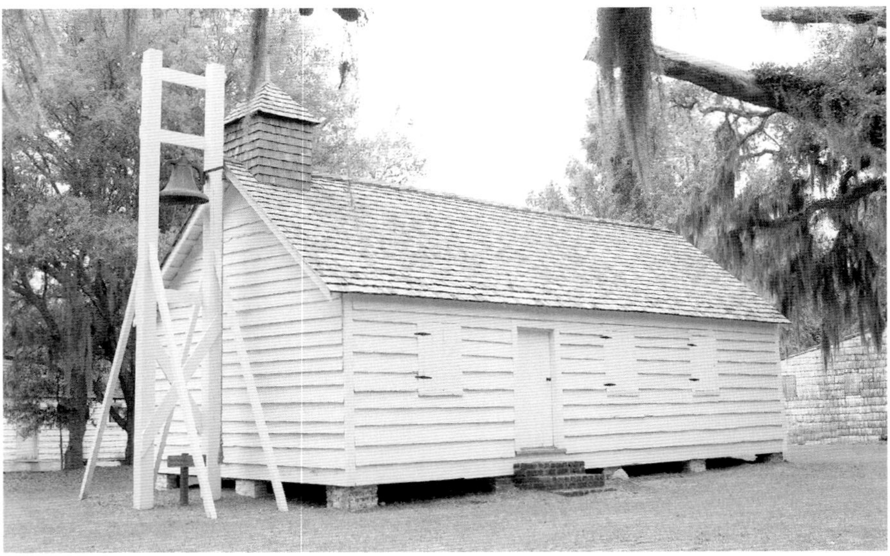

The chapel today, restored to antebellum simplicity.

# The Slave Cabins

The slave cabins were also constructed during the Dr. Francis S. Parker ownership. Each cabin is a duplex with a central fireplace. Because the cabins are separated into two sections and have two entry doors, it is presumed that separate families once lived in each side of the clapboard cabins.

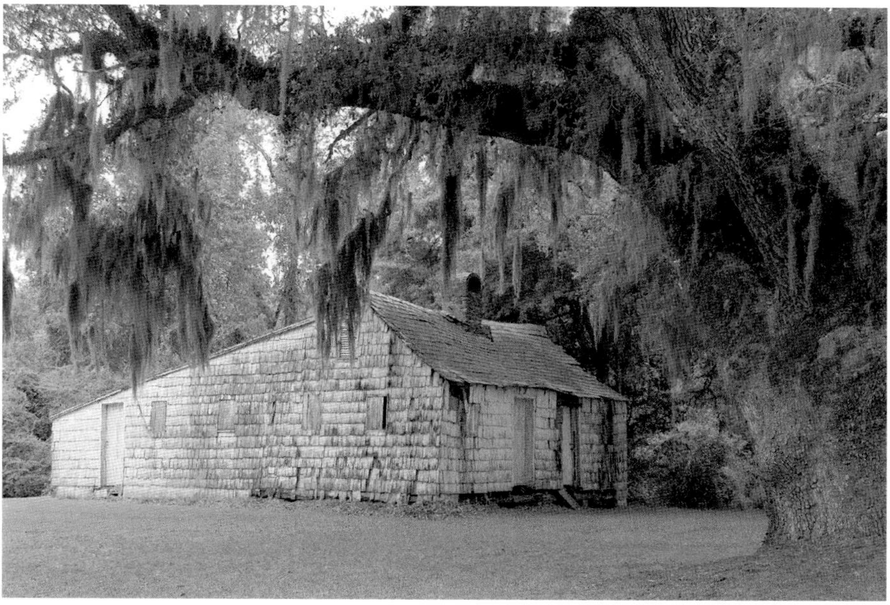

This un-restored cabin, the last one on the south side of the street, was once the schoolhouse that served the Parker slaves. It remained in service for the freedmen, women and children of the village after emancipation.

Each cabin has been altered to include a number of changes. During the Montgomery years, each cabin received running water and electricity, restrooms, a front porch, rear additions, glazed windows, interior walls and ceilings covered in sheetrock. Tar shingles were added to each of the twenty-one- by forty-foot structures.

Currently, the village is under reconstruction. The Parkers and the board of directors of Mansfield Plantation plan to restore the cabins to represent different eras of the plantation's history.

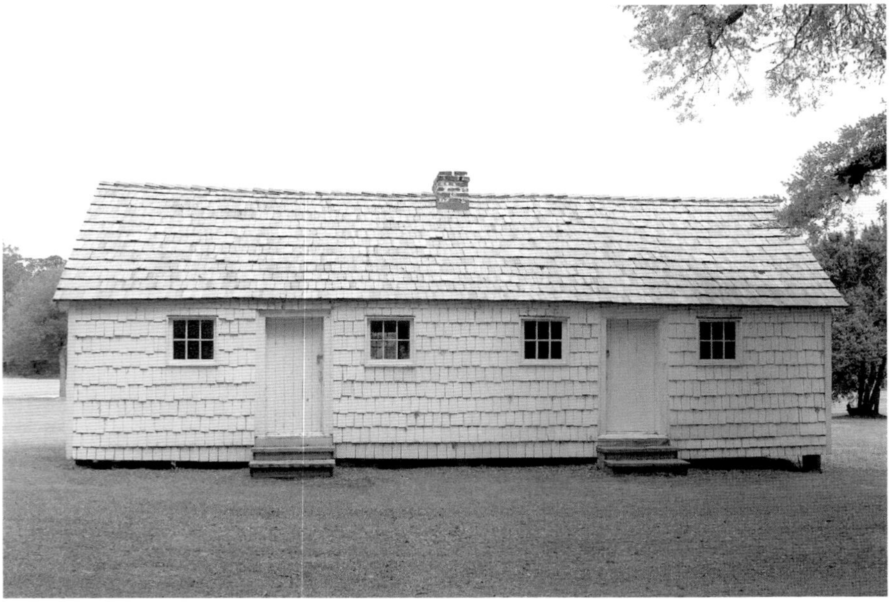

This restored cabin now houses the plantation's museum. These duplex-style cabins were once very common on rice plantations. Each unit provides shelter for two families.

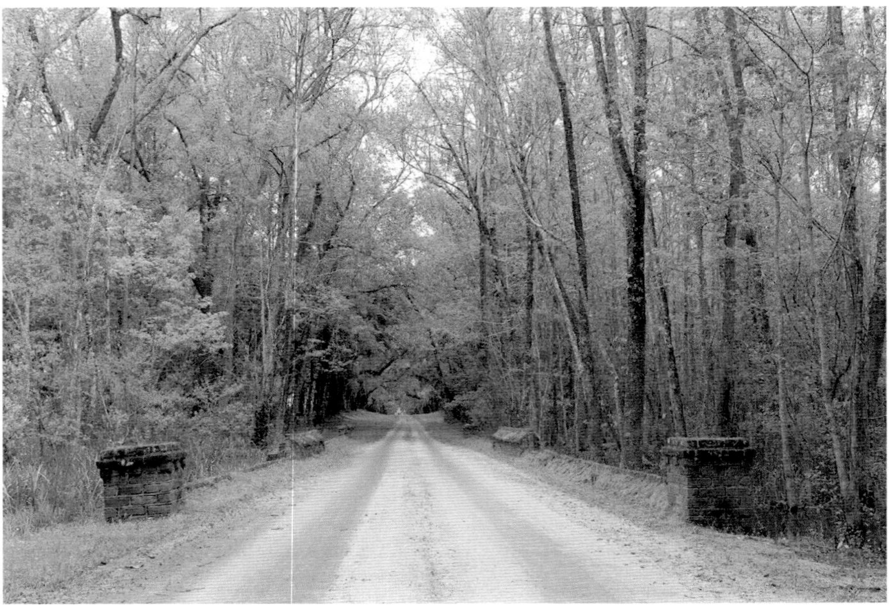

The valley.

## THE VALLEY

The dip in the road at the bridge crossing the swamp, just west of the village, is traditionally referred to as "the valley." This area is the natural drainage route for the uplands of Mansfield Plantation as it connects to hand-dug ditches that drain into the Black River. In the 1930s, Colonel Montgomery's employees bricked the bridge.

## THE AVENUE OF OAKS

The oak alley lines the street from the village to the manor house. The avenue was reportedly planted by the hands of Francis S. and Mary Lance Parker in the 1840s. Planted shortly thereafter in between the oaks, red, white, pink and striped blossoming camellias enhance the driveway.

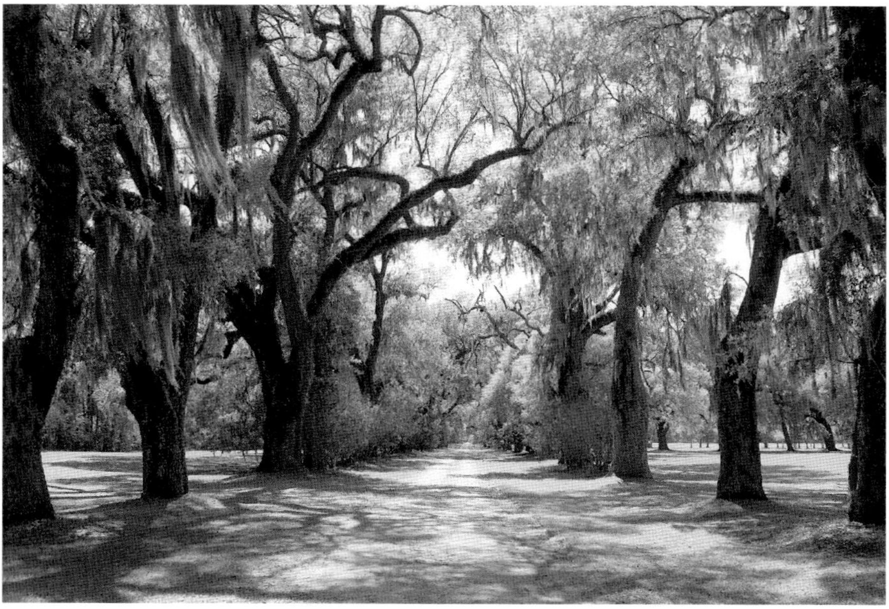

View of the avenue of live oaks from the manor house looking toward the village.

# Notes

## Introduction

1. Smythe et al., *Carolina Low Country*, 149.

## Chapter 1

2. Stampp, *Causes of the Civil War*, 222.
3. Salley, *Introduction of Rice Culture*, 3.
4. Ibid., 11.
5. Littlefield, *Rice and Slaves*, 105, 107, 109. Also see Vernon, *African Americans at Mars Bluff*, 119–23.
6. Salley, *Introduction of Rice Culture*, 4–5
7. Rogers, *History of Georgetown County*, 8.
8. Michie, *Oaks Plantation Revealed*, 5.
9. Salley, *Introduction of Rice Culture*, 6–7.
10. Salley, *Warrants for Land*, 659.
11. Parker, "History of Mansfield."
12. McDowell, *Journals of the Commissioners*, 137, 145, 192, 202, 209, 265.
13. Linder and Thacker, *Historical Atlas*, 453.
14. Cooper, *Statutes at Large*. For examples of some of these construction projects, see pages 69–70, 180–83, 243–45, 335, 390, 431.
15. Rogers, *History of Georgetown County*, 31–36.
16. For the history of the South Carolina Indians, see Corkran, *Carolina Indian Frontier* and Millings, *Red Carolinians*.

17. Salley, *Introduction of Rice Culture*, 16. Charleston exported 100,000 pounds of rice in 1761.
18. Moore, *Abstracts of Wills*, 128.
19. Langley, *South Carolina Deed Abstracts*, vol. 3: 63–64, 75.
20. Ibid., *South Carolina Deed Abstracts* vol. 4: 126–27.
21. Linder and Thacker, *Historical Atlas*, 459.
22. Michie, *Oaks Plantation Revealed*, 5.
23. Morgan, "Organization of the Colonial Rice Trade," 446.
24. Ship Register, 1734–1765, 298, 315–16, South Carolina Archives; Ship Register, 1765–1780, South Carolina Archives.
25. Morgan, "Organization of the Colonial Rice Trade," 435.
26. Ibid., 436–38.
27. Ibid., 435–38.
28. Littlefield, *Rice and Slaves*, 8.
29. Rogers, *History of Georgetown County*, 11.
30. Lauber, *Indian Slavery*, 174.
31. Littlefield, *Rice and Slaves*, 116.
32. Rogers, *History of Georgetown County*, 29.
33. Doar, *Rice and Rice Planting*, 7–10.
34. Gray, *History of Agriculture*, 1024, 1036. These two pages list the value of the exports of both rice and indigo.
35. Linder and Thacker, *Historical Atlas*, 190–91, 211, 379, 395, 584. With the money earned from the sale of the estate, Coachman bought land on Winyah Bay, which later became known as Lucerne and Upton Plantations. A few years later, Coachman bought part of what became Laurel Hill Plantations on the Waccamaw and Oakhampton on Sandy Island in 1762. In 1773, Coachman acquired a 1,065-acre plot on the Pee Dee that later became known as Glenmore when he married the widow Ann Waities Johnston. Two years later, in 1776, he gave 650 acres on the Pee Dee to his son William in "love and good will," which he bought from Anthony White.
36. Parker, "History of Mansfield."
37. Pringle, *Register Book*, 53.
38. Ibid., 23, 25. See also Parker, "History of Mansfield."
39. Linder and Thacker, *Historical Atlas*, 313.
40. Ibid., 313, 453–54, 497–98, 524.
41. Parker "History of Mansfield."
42. Lambert, *South Carolina Loyalists*, 120, 229–30, 261, 265–66; Holcomb, *Marriage, Death, and Estate Notices*, 5.
43. Langley, *South Carolina Deed Abstract* vol. iv: 134.
44. Linder and Thacker, *Historical Atlas*, 313, 453–54, 497–98, 524.
45. Ibid.
46. Ibid.

47. Susannah Man, February 18, 1775, South Caroliniana Library, Collection #2438.

# CHAPTER 2

48. Stampp, *Causes of the Civil War*, 214.
49. Rogers, *History of Georgetown County*, 165.
50. Population Survey, 1790 Census, South Carolina Department of Archives and History.
51. Salley, *President Washington's Tour*, 4.
52. Ibid.
53. Mills, *Statistics of South Carolina*, 559.
54. Ibid., 558.
55. Linder and Thacker, *Historical Atlas*, 461, 528.
56. Talbert and Farish, *Journal of Peter Horry*, 241.
57. Ibid., 243.
58. Holcomb, *Marriage, Death, and Estate Notices*, 20; Holcomb, *Parish Registers*, 93.
59. Holcomb, *Parish Registers*, 93.
60. *Catalogue of the Porcellian Club*, 28.
61. Population Survey, 1810 Census, South Carolina Department of Archives and History; "Journal of General Peter Horry," *South Carolina Historical Magazine* 42, 8, 189.
62. Read-Lance Family Papers, South Caroliniana Library, Collection #10367.
63. Parker, "History of Mansfield"; Johnson and Rosa, *Prince George Winyah*, 39.
64. *Georgetown Gazette*, February 24, 1808.
65. Talbert and Farish, *Journal of Peter Horry*, 243.
66. S213050, Volume 006E, Page 00186, South Carolina Department of Archives and History.
67. Linder and Thacker, *Historical Atlas*, 374, 524. Priscilla Smerden used the money from the sale of the estate to run a boardinghouse in Charleston at 54 Church Street.
68. Talbert and Farish, *Journal of Peter Horry*, 243.
69. *Rectors, Wardens and Vestry*, 2; "The Georgetown Library Society," *South Carolina Historical Magazine* 25, 97; Winyah Indigo Society, *Short History of the Winyah Indigo Society*, 37.
70. "The Georgetown Library Society," *South Carolina Historical Magazine* 25, 100.
71. Talbert and Farish, *Jounal of Peter Horry*, 10, 12, 17, 21, 23, 30, 31, 38, 43, 47, 62, 63, 67, 68, 73, 74, 89, 90, 92.
72. "Journal of General Peter Horry," *South Carolina Historical Magazine* 42, 8, 189.
73. *Winyah Intelligencer*, April 24, 1819.
74. Mills, *Statistics of South Carolina*, 561.
75. Ibid., 570.

76. Ludlum, *Early American Hurricanes*, 52–59; Rogers, *History of Georgetown County*, 225.
77. Population Survey, 1830 Census, South Carolina Department of Archives and History.
78. Ibid.
79. Rogers, *History of Georgetown County*, 224.
80. Sanborn, *Reminiscences of Richard Lathers*, 20–21.
81. Archives in History Series S165015 item 00131, and series S165015 item 01795; "Petition of Francis Kinloch," Legislative System, Slavery Insurrection, South Carolina Department of Archives and History; "Petition of John Coachman," Legislative System, Slavery Insurrection, South Carolina Archives and History.
82. Edgar, *South Carolina Encyclopedia*, 872–73.
83. Thomas, *History of the South Carolina Military Academy*, 5–6.
84. Talbert and Farish, *Journal of Peter Horry*, 172.
85. South Carolina Department of Archives and History, Series S165015, Item 00042.
86. "Journal of Peter Horry," *South Carolina Historical Magazine* 42, 22; J.R. Parker, "History of Mansfield," *Georgetown Times*, March 28, 1941; Lachicotte, *Georgetown Rice Plantations*, 78–81; Holcomb, *Parish Registers*, 93.
87. Johnson and Rosa, *Prince George Winyah*, 22.
88. Mills, *Statistics of South Carolina,* 566.
89. Jones, *Plantation South*, 190.
90. *Georgetown Gazette*, November 16, 1816.
91. Johnson and Rosa, *Prince George Winyah*, 40; *Rectors, Wardens and Vestry*, 1. Thomas, *Episcopal Church in South Carolina*, 537; Holcomb, *Parish Registers*, 2–4, 6, 93.
92. Will of John Man Taylor, South Carolina Department of Archives and History, Series S126061, Item 04931.
93. Simons and Simons Case Records, 1824–1879, South Carolina Historical Society 0431.02(P); Lance Family Legal Papers, 1832–1877, South Carolina Historical Society, 0431.02(L)04.
94. Rogers, *History of Georgetown County*, 206, 363.
95. For an account of both the Methodist and Episcopal Missions to the slaves, see Rogers, *History of Georgetown County*, 349–60.
96. Sanborn, *Reminiscences of Richard Lathers*, 19.
97. Population Survey, 1820 Census, South Carolina Department of Archives and History.
98. South Carolina Department of Archives and History, Series S213003, vol. 005A, page 00036.
99. Tax Records, South Carolina Department of Archives and History, Series S126061, Item 00958.
100. *Southern Agriculturist and Register of Rural Affairs, November* (Charleston, SC: A.E. Miller, November 1829), 504–08.

101. Ibid., *September* 182, 403–06.
102. Olmsted, *Journey in the Seaboard Slave States*, 108–09.
103. Allston, *Analysis of Rice Straw*, 43, 60–61.
104. "Products of Industry," 1860 Census, South Carolina Department of Archives and History.
105. Easterby, *Carolina Rice Plantation*, 25.
106. Thomas, *History of the South Carolina Military Academy*, 537.
107. Dalco, *Historical Account of the Protestant Episcopal Church*, 309.
108. Linder and Thacker, *Historical Atlas*, 560–61.
109. Ibid., 397.
110. Sanborn, *Reminiscences of Richard Lathers,* 20.
111. Ibid., 493, 560, 587, 611.
112. Ibid., 560.
113. www.carolana.com/SC/1800s/antebellum/sc_antebellum_38th_general_assembly_members.html.
114. Population Survey, 1830 Census, South Carolina Department of Archives and History; Parker, "History of Mansfield."
115. Slave Schedule, 1860 Census, South Carolina Department of Archives and History.
116. Read Family Legal Papers, 1841–1879, South Carolina Historical Society, 0308.01(R)01.
117. Wragg Family Papers 431.02 (W) 20, South Carolina Historical Society.
118. Simons and Simons Case Records, South Carolina Historical Society, 431.02(L)04.
119. Linder and Thacker, *Historical Atlas*, 304, 375.
120. Lachicotte, *Georgetown Rice Plantations*, 141, 143.
121. Simons and Simons Case Records, South Carolina Historical Society.
122. See Simons and Simons Case Records, South Carolina Historical Society, 431.02(L)04; Lance Family Legal Papers, 1832–1877, 0431.02(L)04; Simons and Simons Records Relating to Slaves, 1795–1878, South Carolina Historical Society, 0431.10.
123. Linder and Thacker, *Historical Atlas*, 443–44; *Winyah Observer*, November 19, 1844.
124. "Calculations of the Profit on Rice Mill," *Southern Agriculturist and Register of Rural Affairs* (Charleston, SC: A.E. Miller, 1825) 444–48.
125. Population Survey, 1830 Census, South Carolina Department of Archives and History; Population Survey, 1840 Census, South Carolina Department of Archives and History.

## Chapter 3

126. *Pee Dee Times*, "What Constitutes a Gentleman," May 14, 1856.
127. Wragg Family Papers, South Carolina Historical Society 431.02 (W) 2; Linder and Thacker, *Historical Atlas*, 442–43.
128. Ibid.

129. Plantation Slaves of Mansfield, May 1841 Per Equity Records #0023771, Mormon Church Family records for Charleston, SC Bill no. 57.
130. Holcomb, *Parish Registers*, 16–19, 83.
131. Dr. Francis S. Parker Plantation Records: 1841–1862, South Caroliniana Library Collection #1608.
132. Ibid.
133. Ibid.
134. Lawson, *No Heir to Take Its Place*, 4.
135. Clark, *South Carolina*, 210.
136. Winyah Indigo Society, *Short History of the Winyah Indigo Society*, 3–6, 33–37.
137. James Ritchie Sparkman Papers, South Caroliniana Library, 26; South Carolina General Assembly, Petitions, South Carolina Department of Archives and History, #3780–81.
138. Robert F.W. Allston Papers, Minutes Book of the Winyah and All Saints Agricultural Society, South Carolina Historical Society, 34–161.
139. *Constitution of the Winyah and All Saints Agricultural Society*, (Georgetown: Observer Office, 1842), 2; J.R. Easterby, "Constitution of the Winyah and All Saints Agricultural Society," *South Carolina Historical Magazine* 45, 52–54.
140. Clark et al., *History of the Agricultural Society*, 5–6.
141. Minutes Book of the Winyah and All Saints Agricultural Society.
142. Ibid.
143. Legare, *Synopsis of the Culture of Rice*, 80–81
144. Ibid.
145. *New York Daily Times*, "Industrial Exhibition of 1851," October 29, 1851; *Winyah Observer*, November 5, 1851.
146. Allston, *Brook Green Waccamaw*, 28; Pringle, *Chronicles of Chicora Wood*, 36.
147. Minutes Book of the Winyah and All Saints Agricultural Society.
148. Gray, *History of Agriculture*, 1030.
149. Ibid.
150. Easterby, *Carolina Rice Plantation*, 43, 141, 143.
151. Population Survey, 1840 Census, South Carolina Department of Archives and History.
152. Agricultural Survey, 1850 Census, South Carolina Department of Archives and History.
153. Dwellings of White and Free Colored Persons, 1850 Census. Spartanburg contained 3,185 homesteads and Edgefield 3,019. Abbeville (2,391), Anderson (2,440), Barnwell (2,460), Greenville (2,351), Laurens (2,132) and York (2,190) all contained over 2,000.
154. Population Survey, 1850 Census, South Carolina Department of Archives and History.
155. Simons and Simons Records Relating to Slaves, 1795–1878, South Carolina Historical Society, 0431.10.

156. Agricultural Survey, 1850 Census, South Carolina Department of Archives and History; Lachicotte, *Georgetown Rice Plantations*, 138–41.
157. Rogers, *History of Georgetown County*, 337.
158. 1850 Agricultural Survey.
159. Lance Family Papers, 1832–1877, South Carolina Historical Society, 0431.02 (L) 04.
160. Sanborn, *Reminiscences of Richard Lathers*, 19.
161. Will of Mrs. Mary Taylor Lance, South Carolina Historical Society.
162. Lance Family Legal Papers, 1832–1877, South Carolina Historical Society, 0431.02(L)04; South Carolina Historical Society, 0431.02 (L) 04; Simons and Simons Case Records, 1824–1879, 0431.02 (P).
163. *Pee Dee Times*, August 2, 1854; Population Survey, 1830 Census, South Carolina Department of History and Archives.
164. Agricultural Survey, 1860 Census, South Carolina Department Archives and History.
165. Ibid., Slave Schedule, 1860 Census, South Carolina Department of History and Archives; Linder and Thacker, *Historical Atlas*, 375–76; Richardson, *Cases in Equity*, 147–48.
166. Population Survey, 1860 Census, South Carolina Department of History and Archives.
167. 1860 Agricultural Survey, *South Carolina Historical Magazine* 43, 21–22.
168. 1850 and 1860 Agricultural Surveys.
169. Sixth Federal Population Census, 1850, South Carolina Department of Archives and History; Eighth Federal Population Census, 1860, South Carolina Department of Archives and History.
170. Allston, *Analysis of Rice Straw*, 25–27.
171. James Ritchie Sparkman Papers, Secretary's Records of the Planters Club on the Pee Dee, Southern Historical Collection, Collection # 2732; Robert F.W. Allston Papers, Minutes Book of the Winyah and All Saints Agricultural Society, South Carolina Historical Society, Collection #34-161.
172. "Diary of John Berkley Gimball," *South Carolina Historical Magazine* 56, 23, 95.
173. Parker, Records.
174. Thomas Pinckney to William Lowndes, 1826, Pinckney-Lowndes Papers.
175. Linder and Thacker, *Historical Atlas*, 608.
176. 1860 Population Census.
177. Ibid.
178. 1860 Agricultural Survey.
179. 1860 Population Census.
180. Slaves in the Estate of James W. and Emma Gadsden, South Carolina Department of Archives and History; Slaves in the Estate of Mary LaRoche, South Carolina Department of Archives and History, S213050, Volume 006E, Page 00186 and Series S213050, Volume 006E, Page 00185.

181. Parker, Records.
182. Simons and Simons Records Relating to Slaves, 1795–1878, South Carolina Historical Society, 0431.10.
183. South Carolina Department of History and Archives, Series S213050, Volume 006A, Page 00388; Parker, Records.
184. Agricultural Survey, 1860 Census, South Carolina Department of Archives and History.
185. Kirke, *Among the Pines*, 18–19.
186. Jones, *Plantation South*, 206.
187. 1850 and 1860 Population Surveys and Slave Schedules.

## Chapter 4

188. Daughters of the Confederacy, *For Love of a Rebel*, 4–5.
189. Easterby, *Carolina Rice Plantation*, 13.
190. Rogers, *History of Georgetown County*, 310–11.
191. Population Survey, 1850 Census, South Carolina Department of Archives and History; Population Survey, 1860 Census, South Carolina Department of Archives and History.
192. *Winyah Observer*, October 25, 1848; Robert F.W. Allston Papers, Journal of the Hot and Hot Fish Club, South Carolina Historical Society.
193. Malet, *An Errand to the South.*
194. Michie, *Richmond Hill Plantation*, 38.
195. Weston, "Rules on the Rice Estate of P.C. Weston."
196. Easterby, *Carolina Rice Plantation*, 348–49.
197. Ibid.
198. Dr. Francis S. Parker Plantation Records: 1841–1862, South Caroliniana Library Collection # 1608.
199. Weston, "Rules on the Rice Estate," 38–44.
200. Ibid.
201. Easterby, *Carolina Rice Plantation*, 348–49.
202. Parker, Records.
203. Joshua John War Plantation Journal, South Carolina Historical Society, 34–117; Easterby, "South Carolina through New England Eyes," *South Carolina Historical Magazine* 45, (1944): 132; "Recollections of a Visit to the Waccamaw," *Southern Episcopalian* (1857): 293.
204. Morton, *Folk Remedies*, 21–32, 39–90, 93–102, 103–28, 133–43, 153–59, 161–65.

# Chapter 5

205. Singleton, *Archaeology of Slavery and Plantation Life*, 23–27.
206. Michie, *Richmond Hill Plantation*, 121–32.
207. Parker, "History of Mansfield."
208. Singleton, *Archaeology of Slavery and Plantation Life*, 37.
209. Ibid, 255.
210. Fox-Genovese, *Within the Plantation Household*, 47, 111–13, 120–21.
211. Jones, *Plantation South*, 177.
212. Easterby, *Carolina Rice Plantation*, 63.
213. Weston, "Rules on the Rice Estate," 38–44.
214. Ibid.
215. Franklin, "Overseers," *Carolina Planter* 1 (August 1844): 25–30.
216. Dr. Francis S. Parker Plantation Records: 1841–1862, South Caroliniana Library Collection # 1608.
217. Ibid.
218. Population Survey, 1850 Census, South Carolina Department of History and Archives; Population Survey, 1860 Census, South Carolina Department of History and Archives; Slave Schedules, 1850 Census, South Carolina Department of History and Archives; Slave Schedules, 1860 Census, South Carolina Department of History and Archives.
219. Ibid.
220. Scarborough, *The Overseer*, 57.
221. Weston, "Rules on the Rice Estate," 38–44.
222. Ibid.
223. Parker, Records.
224. Olmsted, *Journey in the Seaboard Slave States*, 65–67.
225. Joyner, *Shared Traditions*, 95.
226. Parker, Records.
227. Olmsted, *Journey in the Seaboard Slave States*, 54–57.
228. Scarborough, *The Overseer*, 50–51.
229. Boyle and Fitch, *Georgetown County Slave Narratives*, 41.
230. Parker, Records.
231. Fox-Genovese, *Plantation Household*, 129, 153.
232. Ibid, 155.
233. Childs, *Rice Planter and Sportsman*, 60.
234. James Ritchie Sparkman to Benjamin Allston, Allston Family Papers, South Caroliniana Library.
235. Easterby, *Carolina Rice Plantation*, 92, 117.
236. Ibid, 60–61.
237. Joyner, *Down by the Riverside*, 72.
238. Rowland, "Highly Valuable Information."

239. Olmsted, *Journey in the Seaboard Slave States,* 64–67.
240. Ibid.
241. Benjamin Allston, "Notes on Management of a Southern Rice Estate," *De Bow's Review* 24 (April 1858): 324–26.

## Chapter 6

242. Sass et al., *Carolina Rice Plantation*, 95–96.
243. Boyle and Fitch, *Georgetown County Slave Narratives*, 12–13.
244. Ibid., 32.
245. Easterby, *Carolina Rice Plantation.*
246. Ibid.
247. Ibid.
248. Weston, "Rules on the Rice Estate," 38–44.
249. "Minutes of the Winyah and All Saints Agricultural Society," Robert F. W. Allston Papers, South Carolina Historical Society, 34–161.
250. Sass et al., *Carolina Rice Plantation*, 96.
251. Olmsted, *Journey in the Seaboard Slave States*, 77, 72–73.
252. Weston, "Rules on the Rice Estate," 38–44.
253. Ibid.
254. Wiggins, "Play of Slave Children," 24–28.
255. Hendrix, *Down and Dirty*, 72.
256. Joyner, *Shared Tradition*, 82–85.
257. Ibid., 36–37.
258. Glennie, *Sermons Preached on Plantations*, 11–13.
259. Jones, *Plantation South*, 175.
260. Olmsted, *Journey in the Seaboard Slave States,* 80.
261. Tindall, *South Carolina Negroes*, 5–6, 59, 60–63. John D. and Mary Magill of Richmond Hill Plantation educated some of their slaves very well. One Magill slave, Bruce H. Williams, became an African Methodist preacher shortly after the war and eventually became a state senator in 1876, a post that he held until 1888. The *Charleston News and Courier*, December 1, 1884, stated that Williams had a "dignified and self-respecting demeanor" and that his long term of service earned him the respect of the legislature.
262. Fox-Genovese, *Within the Plantation Household*, 150.
263. Jones, *Plantation South,* 190, 212–23.
264. Malet, *An Errand to the South.*
265. Olmsted, *Journey in the Seaboard Slave States,* 79–80.
266. Joyner, *Shared Traditions*, 52.
267. Ibid.
268. Holcomb, *Parish Registers*; Malet, *An Errand to the South.*
269. Malet, *An Errand to the South.*

270. Ibid.
271. "Recollections of a Visit to the Waccamaw," 293.
272. Weston, Address by Plowden C.J. Weston Before the Citizens of All Saints Parish at Watchesaw Plantation, July 4th 1857.
273. Malet, *An Errand to the South.*
274. Hendrix, *Down and Dirty*, 52–53.
275. Ibid., 53; Trinkley, *Mansfield Plantation African-American Cemetery*, 8.
276. Michie and Boyle, *Survey of Mansfield Plantation*, 20–23.
277. Hendrix, *Down and Dirty*, 74–76.
278. Ibid, 22–23.
279. Ibid, 22.
280. Michie and Boyle, *Survey of Mansfield Plantation*, 21.
281. Ibid, 38–40.
282. Ibid.

# Chapter 7

283. Harper and Hammond, *Proceedings of the Agricultural Convention,* 169. Mr. Dickerson originally read the essay to the State Agricultural Society on November 28, 1843.
284. Pringle, *Chicora Wood*, 67.
285. Ibid., *Woman Rice Planter*, 58.
286. Brewster, *Summer Migrations*, 26.
287. Pringle, *Chicora Wood*, 158–59.
288. Ibid., 159.
289. Joyner, *Down by the Riverside*, 72.
290. Prevost, *Pawleys Island*, 1.
291. Easterby, "South Carolina through New England Eyes," 133.
292. Sanborn, *Reminiscences of Richard Lathers*, 6.
293. Easterby, *Carolina Rice Plantation*, 102.
294. Pringle, *Chicora Wood*, 137.
295. Reverend Alexander Glennie Parish Diary 1832–1859, South Carolina Historical Society, 34-247.
296. Ibid. The original officers were President William Algernon Alston, Vice-President Joshua John Ward, Secretary and Treasurer Reverend Alexander Glennie and directors Thomas Alston, Edward Thomas Heriot and Francis Marion Weston.
297. Prevost, *Pawleys Island*, 33.
298. *Winyah Observer*, January 7, 1843.
299. Bull, *All Saints Church*, 33; Glennie Parish Diary, 34–247. Reverend Glennie often gave several sermons on each Sunday to planter and slave alike. In All Saints Parish, there were thirteen slave chapels; they were located at Woodbourne,

Laurel Hill, Brook Green, Oaks, Litchfield, Waverly, Midway, True Blue, Hagley, Fairfield, Sandy Knoll, Cedar Grove and Mount Arena Plantations.

300. *Acts of the General Assembly of the State of South Carolina* (Columbia , S.C., 1853).

301. Childs, *Rice Planter and Sportsman*, 105; All Saints Church, Waccamaw, South Carolina Historical Society, 45–77.

302. Stoney, "Memoirs of Frederick Adolphus Porcher," 28.

303. Sanborn, *Reminscences of Richard Lathers*, 5.

304. Pringle, *Chicora Wood*, 150–52.

305. Easterby, *Carolina Rice Plantation*, 135–36.

306. Joyner, *Down by the Riverside*, 133–37.

307. Ibid., 127; Pringle, *Chicora Wood*, 152–56.

308. Charleston Taxpayers List for 1859. Dr. Francis S. Parker's Charleston house was valued at $6,500 in 1859 and was staffed year-round by four domestic servants. John Harleston Read II owned two Charleston houses with a combined value of $50,000 and staffed by thirty servants.

309. Rogers, *History of Georgetown County*, 323.

310. Brewster, *Summer Migrations and Resorts*, 1947.

311. Fraser, *Charleston Charleston!*, 284.

312. Pringle, *Chicora Wood*, 214, 219.

313. Ibid., 145–46.

314. Ibid., 143–44.

315. Fraser, *Charleston Charleston!*, 96.

316. Pringle, *Chicora Wood*, 219.

317. Warren, *Fashion Accessories*, 103–05.

318. Ibid., 106–09.

319. Irving, *South Carolina Jockey Club*, 157–58. The fifteen racetracks in the state that composed the circuit were located in Charleston, St. Matthews, Pendleton, Greenville, Barnwell, Newberry, Strawberry, Georgetown, Fulton, Camden, Columbia, Orangeburg, Cherokee Ponds, Limestone Springs and Yorkville.

320. Ibid., 191.

321. Childs, *Rice Planter and Sportsman*, 19.

322. Irving, *South Carolina Jockey Club*, 192–94.

323. Ibid., 196. Read Family Legal Papers, 1841–1879, South Carolina Historical Society, 0308.01(R)01. John H. Read II paid his fees bi-annually at eighty dollars for two years.

324. Fraser, C*harleston Charleston!*, 181.

325. Pringle, *Chicora Wood*, 5–46.

# CHAPTER 8

326. Radley, *Rebel Watchdog*, 4.
327. *Charleston Mercury*, November 8, 1860.
328. Lesser, *Relic of the Lost Cause*, 2–3.
329. Ibid., 21–26.
330. For a complete list of plantations bought, inherited and sold by Governor Robert Francis Withers Allston, see Linder and Thacker, *Historical Atlas*, 8, 25, 26, 31, 102, 104, 121–22, 136–38, 155, 175.
331. "Delegates to the South Carolina Secession Convention of 1860, with a Summary of Data from Manuscript Returns of Schedules 1 and 2 of U.S. Census for 1860," *South Carolina Historical Magazine* 55 (1954): 192–97; Linder and Thacker, *Historical Atlas*, 128–29, 290, 353, 501, 545–46.
332. Linder and Thacker, *Historical Atlas*, 43, 97, 108, 269–70, 429.
333. "Delegates to the South Carolina Secession Convention of 1860," *South Carolina Historical Magazine* 55, 192–97.
334. Ibid., 192–97; Lesser, *Relic of the Lost Cause*, 4.
335. Lesser, *Relic of the Lost Cause*, 8–9.
336. Pringle, *Chicora Wood*, 172–73.
337. Lesser, *Relic of the Lost Cause*, 15.
338. Easterby, *Carolina Rice Plantation*, 418.
339. United Daughters of the Confederacy, *For Love of a Rebel*, 4–5, 46; Parker, "History of Mansfield"; Tower, *Carolinian Goes to War*, 142, 153.
340. Lachicotte, *Georgetown Rice Plantations*, 141; Linder and Thacker, *Historical Atlas*, 375–76, 428.
341. Pringle, *Chicora Wood*, 173.
342. Ibid, 174–75; Rogers, *History of Georgetown County*, 390–92; Tower, *Carolinian Goes to War*, 6–7.
343. *Charleston Mercury*, August 19, 1861.
344. *Charleston News and Courier*, August 27, 1861.
345. Bridell, "That We Should Have a Port," 50.
346. May and Faunt, *South Carolina Secedes*, 55–56.
347. Bridell, "That We Should Have a Port," 50.
348. Rogers, *History of Georgetown County*, 402.
349. Malet, *An Errand to the South*, 18.
350. *Charleston Mercury*, July 26, 1862.
351. Ibid., July 21, 1862.
352. Rogers, *History of Georgetown County*, 402–07.
353. Malet, *An Errand to the South*, 1,4, 8, 9,11, 23.
354. Collins, *Memories of the Southern States*, 29.
355. Schwalm, *A Hard Fight for We*, 296.
356. Ibid., 105.

357. Ibid.
358. Cheves Collection, Middleton Papers, South Carolina Historical Society; Rogers, *History of Georgetown County*, 409.
359. Joyner, *Down by the Riverside*, 55.
360. Michie, *Richmond Hill Plantation*; Simmons, *Defending South Carolina's Coast*, 47.
361. *War of the Rebellion*, series 4, vol. 2, 1900.
362. Ibid., series 1, vol. 28, part II, 1890.
363. Dr. Francis S. Parker to Bragg, December 25, 1863, Bragg Papers, Western Reserve Historical Society.
364. Francis S. Parker to Governor Andrew G. Magrath, December 24, 1864, Series 513004; Governor Andrew Gordon Magrath, Letters Received and Sent, 1864–1865, South Carolina Department of Archives and History.
365. *War of the Rebellion*, series 1, vol. 47, part II, 1900.
366. For a complete history of the Confederate flight from Richmond, see Rembert, *Jefferson Davis and His Cabinet*, 345–68.

## Chapter 9

367. Smythe et al., *Carolina Low Country*, 149–50.
368. Williamson, *After Slavery*, 19–20. Many of the occupation troops, especially in the 102nd, were freed slaves who had been recruited directly into the Union army. A total of 5,462 former South Carolina slaves were inducted into the Union army by the end of the war.
369. Rogers, *History of Georgetown County*, 419–23.
370. Easterby, *Carolina Rice Plantation*, 329.
371. Benjamin Allston to Edward N. Thurston, March 24, 1866, Allston Family Papers.
372. Easterby, *Carolina Rice Plantation*, 206–07.
373. Ibid., 328
374. Ibid., 329.
375. Ibid., 210.
376. Ibid., 209.
377. Sanborn, *Reminiscences of Richard Lathers*, 9.
378. Easterby, *Carolina Rice Plantation*, 205–06.
379. Ibid., 52–53.
380. Abbott, *Freedmen's Bureau in South Carolina*, 56–58. These lands were later returned to the planters after they took the oath of allegiance to the United States.
381. Federal Writer's Project, *Slave Narratives of South Carolina*, vol. 14, 282. Duncan Allston of Midway Plantation was the last of the Georgetown planters to take the oath of allegiance to the United States. Mariah Heywood, an ex-slave of his, claimed that Allston's slaves took pride in the fact that their master was the last to take the oath. Mrs. Genevieve W. Chandler, project

director for Georgetown narratives, stated, "Her head is just a little higher and her backbone just a little stiffer than that of the average colored person because of pride—family pride—in her people—her white people. And as one can see from her testimony, her chief cause for her pardonable snobbery seems to be that her massa was the last man to surrender and 'swear against his swear.'"

382. Easterby, *Carolina Rice Plantation*, 209–11.
383. Rogers, *History of Georgetown County*, 425–29. The plantations that were ran by the Freedmen's Bureau during the 1865 season were Hagley, True Blue, Turkey Hill, Ingleside, Weehaw, Friendfield, Strawberry Hill, Marietta, Forlorn Hope and Clifton. The revenue produced from these plantations was divided equally among the freed slaves and the United States government.
384. Julian A. Selby, Constitution of the State of South Carolina and the Ordinances, Reports and Resolutions Adopted by the Conventions of the People Held in Columbia, SC, September 1865, Columbia, SC 1866.
385. Pringle, *Chicora Wood*, 366.
386. Easterby, *Carolina Rice Plantation*, 215.
387. Capers, *Confederate Military History* V, 421.
388. Rogers, *History of Georgetown County*, 431; Holcomb, *Parish Registers*, 99.
389. Dr. Francis S. Parker Plantation Records, South Caroliniana Library. Dick is listed on the role of slaves as being the driver.
390. Freedmen Contracts, RG 105, National Archives.
391. Sworn statements of Dr. Francis S. Parker and Frank Parker Jr., April 3 and April 11, 1866, RG 98, National Archives.
392. Interview with John Rutledge Parker, November 22, 2012.
393. Sworn statements of Dr. Francis S. Parker and Frank Parker Jr., April 3 and April 11, 1866, RG 98, National Archives.
394. Freedman's Contract; Lachicotte, *Georgetown Rice Plantations*, 141.
395. Read Family Legal Papers, 1820–1876, (431.02 (R) 03), South Carolina Historical Society.
396. Ibid.
397. Parker, "History of Mansfield Plantation."
398. Sanborn, *Reminiscences of Richard Lathers*, 235.
399. James R. Sparkman to Benjamin Allston, November 23, 1866, Allston Family Papers.
400. Lawson, *No Heir to Take Its Place*, 22–23.
401. Rogers, *History of Georgetown County*, 429; Johnson and Rosa, *Prince George Winyah*, 40; Holcomb, *Parish Registers*, 99.
402. Deed Book B, 65, Courthouse, Georgetown, South Carolina.
403. Francis Withers Pickens to Mrs. Adele Petigru Allston, Allston Family Letters.
404. Read Family Legal Papers.
405. Eric Foner, *Short History of Reconstruction*, 121–24.

406. Rogers, *History of Georgetown County*, 436.
407. Oliver H. Kelley to Benjamin Allston, April 18, 1867, Allston Family Papers. For more information on the attempt to lure foreign labor, see John A. Wagener to Benjamin Allston, August 13, 1867, Allston Family Papers. Wagener proposed that the Georgetown planters form a company to import workers to work their fields.
408. Freedman's Contracts, RG 105, National Archives.
409. Rainey's father was a barber who owned two slaves who worked for him in his barbershop. Slave Schedule, 1850 Census, South Carolina Department of Archives and History; Slaves Schedule, 1860 Census, South Carolina Department of Archives and History.
410. Taylor, *Negro in South Carolina*, 197–98.
411. Abbott, *Freedmen's Bureau in South Carolina*, 79. For an account of freedmen education in Georgetown during this period, see Simkins and Woody, *South Carolina During Reconstruction*, 416–17. The School Act of 1870 established the basis for free schools in South Carolina.
412. Rogers, *History of Georgetown County*, 445. Georgetown became a county in 1868.
413. Sanborn, *Reminiscences of Richard Lathers*, 272–73.
414. Federal Writer's Project, *Slave Narratives of South Carolina*, 110.
415. South Carolina Department of Archives and History, Series S192021, Volume 0013, Page 00045.
416. Mary Taylor Lance Parker, South Carolina Historical Society, 43/0175.
417. Dr. Francis S. Parker and Mary Taylor Lance Parker Genealogy; Holcomb, *Parish Registers*, 68, 100.
418. Holcomb, *Parish Registers*, 30–37.
419. Dethoff, *History of the American Rice Industry*, 62. South Carolina produced 44,073 of the country's total 82,499 barrels.
420. South Carolina Rice Plantation Trust, *Reports, Certificates, Proofs and Endorsements*, introduction.
421. Ibid., 2.
422. Benjamin Allston to Charles Richardson Miles, August 17, 1869, Allston Family Papers. Allston claimed that this system was used by both Mr. Hazard on the Santee River and Mr. Ford on the Black River. Some of these notes can be seen at the Georgetown Rice Museum or in Lawson, *No Heir to Take Its Place*, 23. The notes that can be viewed either at the museum or in the pamphlet were issued at Estherville Plantation and were redeemable in various denominations ranging from ten cents to two dollars.
423. Agricultural Survey, 1860 Census; South Carolina Department of Archives and History; Agricultural Survey, 1870 Census, South Carolina Department of Archives and History.
424. Ibid.
425. Ibid.

426. Linder and Thacker, *Historical Atlas*, 552; Holcomb, *Parish Registers*, 100.
427. Agricultural Survey, 1870.
428. South Carolina Rice Plantation Trust, *Reports, Certificates, Proofs and Endorsements*, 2.
429. Ibid.
430. Lawson, *No Heir to Take Its Place*, 25.
431. The task system required that each worker perform a certain amount of work daily. Usually, each adult male was assigned a half acre of land to work each day. The system differed from the gang labor that was employed on the cotton and tobacco plantations that allowed the workers to work together in the fields.
432. *Georgetown Rice Milling Company*, (Georgetown, SC: Georgetown Rice Milling Company, 1880), 12. The chart was duplicated and is also published in Dethloff, *History of the American Rice Industry*, 64. In 1876, Louisiana produced 64,885 barrels, South Carolina produced 45,591 barrels and Georgia produced 25,199, for a total of 135,653 barrels.
433. Childs, *Rice Planter and Sportsman*, 132–33.
434. Rogers, *History of Georgetown County*, 461. The club's uniform was a gray single-breasted frock coat edged with green cord and faced with green. It also had a single row of palmetto buttons down the front and three smaller buttons on each sleeve. The uniform also required that a feather be worn in the cap.
435. *Charleston News and Courier*, "Hampton in Georgetown," November 3, 1876.

## Chapter 10

436. Ayers, *Promise of the New South*, viii.
437. Tindall, *South Carolina Negroes*, 63.
438. Hollis, *College to University*, 71–73.
439. *South Carolina, Resources and Population, Institutions and Industries* (Charleston, SC, 1883), 70.
440. Lawson, *No Heir to Take Its Place*, 25.
441. Doyle, *A View of Our Past*, 73–75. The company was formed with a capital stock of $60,000 in 1879 and traded primarily with New York and Charleston during its short period of operations.
442. *Georgetown Rice Milling Company*, 6.
443. Lawson, *No Heir to Take Its Place*, 25. The freshets were most destructive to the Pee Dee River plantations because the Pee Dee is the natural drainage route for the mountains of the Carolinas.
444. Lecture by College of Charleston professor Dr. Scott Poole, June 2010.
445. *South Carolina Historical and Genealogical Magazine* 31 (1930): 205; Dr. Parker, Dr. Francis S. and Mary Lance Parker Genealogy; Holcomb, *Parish Registers*, 100, 102, 68.

446. Population Survey, 1880 Census, South Carolina Department of Archives and History.
447. Agricultural Survey, 1880 Census, South Carolina Department of Archives and History
448. Parker, Dr. Francis S. and Mary Lance Parker Genealogy; Holcomb, *Parish Registers*, 36–37, 69.
449. Population Survey, 1880 Census; Holcomb, *Parish Registers*, 36-39.
450. 1880 Agricultural Survey.
451. Johnson and Rosa, *Prince George Winyah*, 37, 40; Deed Book K, 339–42, Courthouse, Georgetown, South Carolina; *Rectors, Warden and Vestry*, 6; Holcomb, *Parish Registers*, 83, 102–04.
452. Johnson and Rosa, *Prince George Winyah*, 303; *Childhood Diary of William Wigg Hazzard*, 1891.
453. Agricultural Surveys 1860–1880.
454. Rogers, *History of Georgetown County*, 468–69.
455. Daniel, *Breaking the Land*, 40. Daniel contends that the influx of midwestern farmers was due to high interest rates, unfavorable climate and crop damage due to grasshoppers.
456. Ayers, *Promise of the New South*, 193.
457. Louisiana did have an antebellum rice industry. It is often overlooked due to its small export value as compared to its sugarcane industry; however, it did exist.
458. Dethloff, *History of the American Rice Industry*, 63. Instead of using floodgates, which needed to be repaired often, the Louisiana rice culture used water pumps to flood the rice fields.
459. Daniel, *Breaking the Land*, 46. The estimate of breaking fifty acres a day was based on running the tractors twenty-four hours a day.
460. Ibid., 83.
461. Ibid., 66. In 1909, South Carolina produced 541,570 bushels of rice while, Arkansas produced 1,282,830, Texas produced 8,991,745 and Louisiana produced 10,839,973. In 1919, South Carolina produced 122,465 bushels of rice, while Arkansas produced 6,797,126, Texas produced 5,306,369 and Louisiana produced 16,011,607.
462. Agricultural Surveys of 1860, 1870, 1880, 1890, 1900 and 1910.
463. Linder and Thacker, *Historical Atlas*, 552. Holcomb, *Parish Registers*, 100.
464. Poole lecture and inscriptions on Georgetown monument.
465. Doyle, A *View of Our Past*, 86–87. The mill brought much opportunity to the town. To accommodate the millworkers' needs, new streets were laid out, sewer and water systems were built and a hotel and mercantile stores were constructed. Block after block of housing was built for laborers and executives. The mill produced Georgetown's biggest expansion since it was first laid out in 1729. The mill employed five thousand workers during its height of production.

466. Bridwell, *Gem of the Atlantic Seaboard*, 32.
467. Ibid., 33.
468. Doar, *Rice and Rice Planting*, 27.
469. Pringle, *A Woman Rice Planter*, 155.
470. Ibid., 263–67.
471. Ibid., 5, 141, 377. Elizabeth Waties Allston Pringle was the last member of the Allston/Alston lineage to plant rice in Georgetown. Her family had carried on the tradition for over 150 years and produced two governors of South Carolina, one of whom, Robert Francis Withers Allston, was her father.
472. Parker, "History of Mansfield."
473. Population Survey, 1910 Census, South Carolina Department of Archives and History.
474. Johnson and Rosa, *Prince George Winyah*.
475. Deed Book E-1, 308-09, Courthouse, Georgetown South Carolina; Plat Book D, 11, Courthouse, Georgetown, South Carolina.
476. Deed Book E-1, 308-09, Courthouse, Georgetown South Carolina; Plat Book D, 11, Courthouse, Georgetown, South Carolina; 1920 Population Census of South Carolina.
477. Parker, "History of Mansfield."
478. Allston, *Brookgreen Waccamaw*, 6. "Carolina Rice" was first grown by Joshua John Ward at Brook Green Plantation in the late 1840s and was first exported in 1851. In that same year, E.T. Heriot won a prize at London's World Fair for his sample of "Carolina Rice." During the 1850s, the height of the rice culture, Georgetown's planters proudly boasted that "Carolina Rice" was their contribution to Western Civilization.
479. Doar, *Rice and Rice Planting*, 41. In 1920, a combined total of only five hundred acres of rice was planted on all nine of South Carolina's rivers.

## Epilogue

480. Deed Book L-2, 47, Courthouse, Georgetown, South Carolina.
481. Lachicotte, *Georgetown Rice Plantations*, 82–84.
482. Deed Book 94, 363–68, Courthouse, Georgetown, South Carolina.
483. Farm Service Agency Report, A.S.C.S. 578, Clemson University Extension Service, Georgetown, South Carolina, 1983-88.
484. United States Department of the Interior, National Park Service, National Register of Historic Places Inventory, Nomination Form, July 5, 1977, entered December 6, 1977.
485. Deed Book 621, 326-29, Courthouse, Georgetown, South Carolina.

# Bibliography

## Primary Sources

### *Manuscripts*


Benjamin Allston to Edward N. Thurston, March 24, 1866. Allston Family Papers. South Caroliniana Library.

Cheves Collection. Middleton Papers. South Carolina Historical Society.

A Childhood Diary of William Wigg Hazzard, January–May, 1891. Personal possession.

Dr. Francis S. Parker Plantation Records. MS 1608. South Caroliniana Library.

Dr. Francis S. Parker to General Braxton Bragg. Braxton Bragg Papers. Western Reserve Historical Society.

Dr. Francis S. Parker to Governor Andrew G. Magrath. Governor Andrew Gordon Magrath Letters Received and Sent, 1864–1865. MS Series 513004. South Carolina Department of Archives and History.

Dr. James Ritchie Sparkman Papers: The Secretary's Records of the Planters' Club on the Pee Dee. MS 2732, Vol. 2. Southern Historical Collection.

Farm Service Agency Report, 1983–88. Clemson University Extension Service, Georgetown.

Francis Withers Pickens to Mrs. Adele Petigru Allston. Allston Family Papers. South Caroliniana Library.

James Ritchie Sparkman Papers. MS 2606. South Carolina Department of Archives and History.

James Ritchie Sparkman to Benjamin Allston. Allston Family Papers, MS 29. South Caroliniana Library.

John Man Taylor Petition against the Election of Thomas P. Davis. MS Series S165015, Item 00042. South Carolina Department of Archives and History.

Joshua John Ward Plantation Journal. MS 34-117. South Carolina Historical Society.

Lance Family Legal Papers, 1832–1877. MS 0431.02(L)04. South Carolina Historical Society.

Lance Family Papers. MS Series S213003, Vol. 005A. South Carolina Department of Archives and History.

Mary Taylor Lance Parker. MS 43/0175. South Carolina Historical Society.

Militia Muster List. MS Series S192021, Volume 0013, Page 00045. South Carolina Department of Archives and History.

"Petition of Francis Kinloch," Legislative System, Slavery Insurrection. MS S165015, Item 00131. South Carolina Department of Archives and History.

"Petition of John Coachman," Legislative System, Slavery Insurrection. MS S165015, Item 01795. South Carolina Department of Archives and History.

Plantation Slaves of Mansfield Plantation. MS 57. Mormon Church Family Records, Charleston, South Carolina.

Read Family Legal Papers, 1841–1879. MS 0308.01(R) 01. South Carolina Historical Society.

Read-Lance Family Papers. MS 10367. South Caroliniana Library.

Reverend Alexander Glennie Parish Diary. MS 34-247. South Carolina Historical Society.

Robert F.W. Allston Papers. Journal of the Hot and Hot Fish Club. MS 12-4-12. South Carolina Historical Society.

———. Minutes Book of the Winyah and All Saints Agricultural Society. MS 34-161. South Carolina Historical Society.

Ship Register, 1734–1765. South Carolina Department of Archives and History.

Ship Register, 1765–1780. South Carolina Department of Archives and History.

Simons and Simons Case Records, 1824–1879. MS 0431.02(P). South Carolina Historical Society.

Simons and Simons Records Relating to Slaves, 1795–1878. MS 0431.10. South Carolina Historical Society.

Slaves in the Estate of James W. and Emma Gadsden. MS S213050, Vol. 006E. South Carolina Department of Archives and History

Slaves in the Estate of Mary LaRoche. MS 006E, Page 00186. South Carolina Department of Archives and History.

South Carolina General Assembly Petitions. MS 3780–81. South Carolina Department of Archives and History.

Susannah Man. Receipt for Sale of Rice to Wragg and Smith Company, February 18, 1775. MS 2438. South Caroliniana Library.

Thomas Pinckney to William Lowndes, 1826. Harriott Ravenel Collections. Pinckney-Lowndes Papers. South Carolina Historical Society.

Will of John Man Taylor. MS Series S126061, Item 04931. South Carolina Department of Archives and History.
Will of Mrs. Mary Lance Taylor. South Carolina Historical Society.
Wragg Family Papers. MS 431.02(W). South Carolina Historical Society.

## *Government Documents*

Agricultural Survey, 1850 Census. South Carolina Department of Archives and History.
Agricultural Survey, 1860 Census. South Carolina Department of Archives and History.
Agricultural Survey, 1880 Census. South Carolina Department of Archives and History.
Deed Book B, 65. Georgetown County Courthouse, South Carolina.
Deed Book E-1, 308-09, Plat Book D 11. Georgetown County Courthouse, South Carolina.
Deed Book K, 339-42. Georgetown County Courthouse, South Carolina.
Deed Book L-2, 47. Georgetown County Courthouse, South Carolina.
Deed Book 94, 363-68. Georgetown County Courthouse, South Carolina.
Deed Book 621, 326-29. Georgetown County Courthouse, South Carolina.
Dwellings of White and Free Colored Persons, 1850 Census. South Carolina Archives and History.
Freedman's Contract. MS RG 105. National Archives.
Population Survey, 1790 Census. South Carolina Department of Archives and History.
Population Survey, 1800 Census. South Carolina Department of Archives and History.
Population Survey, 1810 Census. South Carolina Department of Archives and History.
Population Survey, 1820 Census. South Carolina Department of Archives and History.
Population Survey, 1830 Census. South Carolina Department of Archives and History.
Population Survey, 1840 Census. South Carolina Department of Archives and History.
Population Survey, 1850 Census. South Carolina Department of Archives and History.
Population Survey, 1860 Census. South Carolina Department of Archives and History.
Population Survey, 1880 Census. South Carolina Department of Archives and History.

Population Survey, 1910 Census. South Carolina Department of Archives and History.

Slave Schedule, 1840 Census. South Carolina Department of Archives and History.

Slave Schedule, 1850 Census. South Carolina Department of Archives and History.

Slave Schedule, 1860 Census. South Carolina Department of Archives and History.

Sworn Statement of Dr. Francis S. Parker, April 3 and April 11, 1866. MS RG 98. National Archives.

Sworn Statement of Frank Parker, April 3 and April 11. MS RG 98. National Archives.

## *Newspapers*

*Charleston Mercury*, November 8 and 9, 1860; August 19, 1861; July 21 and 26, 1862.

*Charleston News and Courier*, August 27, 1861; November 3, 1876; December 1, 1884.

*Georgetown Times*, March 28, 1941.

*New York Daily Times*, October 29, 1851.

*Pee Dee Times*, August 2, 1854; January 31, 1855; May 14, 1856.

*Winyah Intelligencer*, April 24, 1819.

*Winyah Observer*, January 7, 1843; November 5, 1851.

## *Books and Pamphlets*

*Acts of General Assembly of the State of South Carolina, Passed in December 1853*. Columbia, SC: R.W. Gibbs & Co., 1853.

Allston, Robert F.W. *Analysis of Rice Straw, Chaff & Supplement to the Proceedings of the State Agricultural Society of South Carolina*. Columbia, SC: Summer & Carroll, 1847.

Boyle, Christopher C., and James A. Fitch. *Georgetown County Slave Narratives*. Georgetown, SC: Seaport Press, 1997.

Capers, Ellison. *Confederate Military History*. Vol. 5. Nashville, TN: Blue and Gray Press, n.d.

*Catalogue of the Porcellian Club of Harvard University*. Cambridge, MA: Harvard University, 1906.

Childs, Arney R. *Rice Planter and Sportsman: The Recollections of J. Motte Alston, 1821–1909*. Columbia: University of South Carolina Press, 1953.

Clark, Thomas Dionysius. *South Carolina: The Grand Tour, 1780–1865*. Columbia: University of South Carolina Press, 1973.

Clark, W.A., ed. *History of the Agricultural Society of South Carolina*. Columbia, SC: R.L. Ryan, 1916.

Collins, Elizabeth. *Memories of the Southern States*. Taunton, MA: Barnicott, 1865.

Cooper, Thomas, ed. *Statutes at Large of South Carolina*. Vol. 3. Columbia, SC: A.S. Johnston Publishing, 1838.

Easterby, J.H., ed. *The South Carolina Rice Plantation as Revealed in the Papers of Robert F.W. Allston*. Chicago: University of Chicago Press, 1945.

Federal Writers Project. *Slave Narratives of South Carolina*. Vol. 14. East Claire Shores, MI: Scholarly Press, 1976.

*Georgetown Rice Milling Company*. Charleston, SC: Lucas and Richardson, 1880.

Harper, William, and James Henry Hammond. *The Proceedings of the Agricultural Convention and of the State Agricultural Society of South Carolina, from 1839 to 1845*. Columbia, SC: Summer & Carroll, 1846.

Holcomb, Brent. *Marriage, Death, and Estate Notices from Georgetown, S.C. Newspapers, 1791–1861*. Easley, SC: Southern Historical Press, 1979.

———. *Parish Registers of Prince George Winyah Church, Georgetown, South Carolina, 1815–1936*. Columbia, SC: SCMAR, 1996.

Irving, John Beaufain. *The South Carolina Jockey Club*. Charleston, SC: Walker and Evans, 1857.

Johnson, Louise R., and Julia Rosa. *Prince George Winyah: Inscriptions from the Churchyard, 1773–1932*. Georgetown, SC: 1971.

Jones, Katharine M., ed. *The Plantation South*. Indianapolis, IN: Bobbs-Merrill, 1957.

Kirke, Edmund. *Among the Pines: Or, South in Secession-Time*. New York: Charles T. Evans, 1862.

Lance, Maurice Harvey. *An Address Delivered in St. Paul's Church, Radcliffeborough, before the Charleston Protestant Episcopal Sunday School, on Their Twelfth Anniversary, May 24th, Being the Tuesday in Whitsun-Week, 1831*. Charleston, SC: A. E. Miller, 1831.

Langley, Clara A., ed. *South Carolina Deed Abstracts: 1755–1767*. Vol. 3. Easley, SC: Southern Historical Press, 1984.

———. *South Carolina Deed Abstracts: 1767–1773*. Vol. 6. Easley, SC: Southern Historical Press, 1984.

Malet, William Wyndham. *An Errand to the South in the Summer of 1862*. London: R. Bentley, 1863.

McDowell, William L., Jr. *Journals of the Commissioners of the Indian Trade: September 20, 1710–August 29, 1718*. Columbia: South Carolina Department of Archives, 1955.

Mills, Robert. *Statistics of South Carolina, including a View of Its Natural, Civil, and Military History, General and Particular*. Charleston, SC: Hurlbut and Lloyd, 1826.

Olmsted, Frederick Law, and William P. Trent. *A Journey in the Seaboard Slave States in the Years 1853–1854, with Remarks on Their Economy*. New York: G.P. Putnam's Sons, 1904.

Pringle, Elizabeth W. Allston. *Chronicles of Chicora Wood*. Boston: Christopher Pub. House, 1940.

———. *The Register Book for the Parish, Prince Frederick, Winyah*. Baltimore: Williams & Wilkins Company, 1916.

———. *A Woman Rice Planter*. Columbia: University of South Carolina Press, 1994.

Salley, A.S. *Warrants for Land in South Carolina*. Columbia: Historical Commission of South Carolina, 1910.

Sanborn, Alvan F., ed. *Reminiscences of Richard Lathers: Sixty Years of a Busy Life in South Carolina, Massachusetts and New York*. New York: Grafton Press, 1907.

Selby, Julian A., ed. *Constitution of the State of South Carolina and the Ordinances, Reports and Resolutions Adopted by the Convention of the People, Held in Columbia, S.C., September, 1865*. Columbia, SC: Julian A. Selby, 1866.

South Carolina Department of Agriculture. *South Carolina: Resources and Population, Institutions and Industry*. Charleston, SC: Walker, Evans & Cogswell, 1883.

South Carolina Rice Plantation Trust. *South Carolina Rice Plantation Trust: Reports, Certificates, Proofs and Endorsements*. Charleston, SC: South Carolina Rice Plantation Trust, 1873.

U.S. War Department. *The War of the Rebellion: A Compilation of the Official Records of the Union and Confederate Armies*. Washington, D.C.: Government Printing Office, 1900.

Weston, Plowden C.J. *Address by Plowden C.J. Weston before the Citizens of All Saints Parish at Wachesaw Plantation, July 4th, 1857*. N.p., 1857.

Winyah and All Saints Agricultural Society. *Constitution of the Winyah and All Saints Agricultural Society*. Georgetown, SC: Winyah Observer Printing Office, 1842.

## *Journals*

Allston, Benjamin. "Notes on Management of a Southern Rice Estate." *DeBow's Review* (April 1858).

"Calculations of the Profit on Rice Mill." *Southern Agriculturalist and Register of Rural Affairs* (1825).

"Delegates to the South Carolina Secession Convention of 1860, with a Summary of Data from Manuscript Return Schedules 1 and 2 or U.S. Census for 1860." *South Carolina Historical Magazine* 55 (1954).

"Diary of John Berkley Grimball." *South Carolina Historical Magazine* 56 (1957).

"Dr. Francis S. Parker and Mary Lance Parker Genealogy." *South Carolina Historical and Genealogical Magazine* 31 (1930).

Easterby, J.H., ed. "Constitution of the Winyah and All Saints Agricultural Society." *South Carolina Historical Magazine* 45 (1944).

———. "South Carolina through New England Eyes: Almira Coffin's Visit to the Low Country in 1851." *South Carolina Historical Magazine* 45 (1944).

Franklin. "Overseers." *Carolina Planter* (August 1844).

"The Georgetown Library Society." *South Carolina Historical and Genealogical Magazine* 25, no. 2 (April 1924): 94–100.

"Journal of Peter Horry." *South Carolina Historical Magazine* 42 (1941).

Legare, J.D. "Rice Mills." *Southern Agriculturalist and Register of Rural Affairs* (November 1829).

———. "Synopsis of the Culture of Rice on Black River." *Southern Agriculturalist and Register of Rural Affairs* (1841).
Morgan, Kenneth. "The Organization of the Colonial American Rice Trade." *William and Mary Quarterly* 52, no. 3 (1995): 433–52.
"Recollections of a Visit to the Waccamaw." *Southern Episcopalian* 6 (1857).
Stone, Samuel G., ed. "Memoirs of Frederick Adolphus Porcher." *South Carolina Historical Magazine* 46 (1945).
Weston, Plowden C.J. "Rules on the Rice Estate of P.C. Weston, South Carolina, 1856." *DeBow's Review* 21 (January 1857).
Wiggins, David K. "The Play of Slave Children in the Plantation Communities of the Old South, 1820–1860." *Journal of Sports History* 7, no. 2 (Summer 1980).

### *Reports*

National Register of Historic Places. "Mansfield Plantation Nomination Form." July 5, 1977.
Trinkley, Michael. "Mansfield Plantation African-American Cemetery." Columbia, SC: Chicora Foundation, 2009.

## SECONDARY SOURCES

### *Interviews and Lectures*

Parker, John Rutledge. Interview with the author. November 22, 2012.
Pool, Scott. "The Lost Cause." Lecture at Horry County School District Office, Conway, SC, June 2010.

### *Books*

Abbott, Martin Linton. *The Freedmen's Bureau in South Carolina, 1865–1872*. Chapel Hill: University of North Carolina, 1967.
Allston, Susan Lowndes. *Brookgreen, Waccamaw, in the Carolina Low Country*. Charleston, SC: Walker, Evans & Cogswell, 1935.
Arthur Manigault Chapter of the United Daughters of the Confederacy, Georgetown, SC. *For Love of a Rebel*. Charleston, SC: Walker, Evans & Cogswell, 1964.
Ayers, Edward L. *The Promise of the New South: Life after Reconstruction*. New York: Oxford University Press, 1992.
Brewster, Lawrence Fay. *Summer Migrations and Resorts of South Carolina Low-Country Planters*. Durham, NC: Duke University Press, 1947.
Bridwell, Ronald. *Gem of the Atlantic Seaboard*. Georgetown, SC: Georgetown Times, 1991.

———. *"That We Should Have a Port:" A History of the Port of Georgetown, South Carolina, 1732–1865*. Georgetown, SC: Georgetown Times, 1982.

Bull, Henry DeSaussure. *All Saints Church, Waccamaw: The Parish, the Place, the People, 1739–1968*. Georgetown, SC: Winyah Press, 1968.

Corkran, David H. *The Carolina Indian Frontier*. Columbia: University of South Carolina Press, 1973.

Dalco, Frederick. *Historical Account of the Protestant Episcopal Church in South Carolina*. Charleston, SC: A.E. Miller, 1820.

Daniel, Pete. *Breaking the Land: The Transformation of Cotton, Tobacco, and Rice Cultures since 1880*. Urbana: University of Illinois Press, 1985.

Dethoff, Henry. *A History of the American Rice Industry, 1685–1985*. College Station: Texas A&M University Press, 1988.

Doar, David, A.S. Salley, and Theodore D. Ravenel. *Rice and Rice Planting in the South Carolina Low Country*. Charleston, SC: Charleston Museum, 1936.

Doyle, Patricia D. *A View of Our Past: The Morgan Photographic Collection Depicting Georgetown, South Carolina, 1890–1915*. Georgetown, SC: Georgetown Library System, 1973.

Edgar, Walter B. *The South Carolina Encyclopedia*. Columbia, SC: University of South Carolina Press, 2006.

Foner, Eric. *A Short History of Reconstruction, 1863–1877*. New York: Harper & Row, 1990.

Fox-Genovese, Elizabeth. *Within the Plantation Household: Black and White Women of the Old South*. Chapel Hill: University of North Carolina Press, 1988.

Fraser, Walter J. *Charleston! Charleston!: The History of a Southern City*. Columbia: University of South Carolina Press, 1989.

Gray, L.C., and Esther Katherine Thompson. *History of Agriculture in the Southern United States to 1860*. Vol. 2. Gloucester, MA: Peter Smith, 1958.

Hendrix, Michael Patrick. *Down and Dirty: Archaeology of the South Carolina Lowcountry*. Charleston, SC: The History Press, 2006.

Hollis, D.W. *College to University*. Columbia: University of South Carolina Press, 1956.

Joyner, Charles W. *Down by the Riverside: A South Carolina Slave Community*. Urbana: University of Illinois Press, 1984.

———. *Shared Traditions: Southern History and Folk Culture*. Urbana: University of Illinois Press, 1999.

Lachicotte, Alberta Morel. *Georgetown Rice Plantations*. Georgetown, SC: Georgetown Historical Society, 1993.

Lambert, Robert Stansbury. *South Carolina Loyalists in the American Revolution*. Columbia: University of South Carolina Press, 1987.

Lauber, Almon Wheeler. *Indian Slavery in Colonial Times within the Present Limits of the United States*. Williamstown, MA: Corner House, 1979.

Lawson, Dennis T. *No Heir to Take Its Place: The Story of Rice in Georgetown County, South Carolina*. Georgetown, SC: Rice Museum, 1972.

Lesser, Charles H. *Relic of the Lost Cause: The Story of South Carolina's Ordinance of Secession*. Columbia: South Carolina Department of Archives and History, 1990.

Linder, Suzanne Cameron, and Marta L. Thacker. *Historical Atlas of the Rice Plantations of Georgetown County and the Santee River*. Columbia: Published by the South Carolina Department of Archives and History for the Historic Ricefields Association, 2001.

Littlefield, Daniel C. *Rice and Slaves: Ethnicity and the Slave Trade in Colonial South Carolina*. Baton Rouge: Louisiana State University Press, 1981.

Ludlum, David M. *Early American Hurricanes, 1492–1870*. Boston: American Meteorological Society, 1963.

May, John Amasa, and Joan Reynolds Faunt. *South Carolina Secedes*. Columbia: University of South Carolina Press, 1960.

Michie, James L., and Christopher C. Boyle. *An Intensive Archaeological Survey of Mansfield Plantation, Georgetown County, South Carolina*. Conway, SC: Waccamaw Center for Cultural and Historical Studies, 1996.

Michie, James L. *The Oaks Plantation Revealed: An Archaeological Survey of the Home of Joseph and Theodosia Burr Alston, Brookgreen Gardens, Georgetown County, South Carolina*. Conway, SC: Waccamaw Center for Historical and Cultural Studies, 1993.

———. *Richmond Hill Plantation, 1810–1868: The Discovery of Antebellum Life on a Waccamaw Rice Plantation*. Spartanburg, SC: Reprint Co., 1990.

Milling, Chapman James. *Red Carolinians*. Columbia: University of South Carolina Press, 1969.

Morton, Julia F. *Folk Remedies of the Low Country*. Miami, FL: E.A. Seeman Publishing, 1974.

Obernuefemann, Kelly. *The Hands of Hampton: Slavery on a St. James Santee Rice Plantation*. McClellanville, SC: Hampton Plantation State Historic Site, 2000.

Patrick, Rembert W. *Jefferson Davis and His Cabinet*. Baton Rouge: Louisiana State University Press, 1944.

Prevost, Charlotte Kaminski, and Effie Leland Wilder. *Pawleys Island: An Historical Sketch of the Blessed Isle and Its Environs*. Columbia, SC: State Print Co., 1972.

Radley, Kenneth. *Rebel Watchdog: The Confederate States Army Provost Guard*. Baton Rouge: Louisiana State University Press, 1989.

Richardson, James Sanders Guignard. *Reports of Cases in Equity, Argued and Determined in the Court of Appeals, South Carolina, November and December Term, 1858*. Columbia: A.S. Johnston, 1868.

Rogers, George C. *The History of Georgetown County, South Carolina*. Columbia: University of South Carolina Press, 1970.

Salley, A.S. *The Introduction of Rice Culture into South Carolina*. Columbia, SC: Printed for the Commission by the State, 1919.

———. *President Washington's Tour through South Carolina in 1791*. Columbia, SC: Printed for the Commission by the State, 1932.

Sass, Herbert Ravenel, D.E. Huger Smith and Alice R.H. Smith. *A Carolina Rice Plantation of the Fifties: 30 Paintings in Water-Color*. New York: W. Morrow and Co., 1936.

Scarborough, William Kauffman. *The Overseer: Plantation Management in the Old South*. Baton Rouge: Louisiana State University Press, 1966.

Schwalm, Leslie A. *A Hard Fight for We: Women's Transition from Slavery to Freedom in South Carolina*. Urbana: University of Illinois Press, 1997.

Simkins, Francis Butler, and Robert H. Woody. *South Carolina during Reconstruction*. Chapel Hill: University of North Carolina Press, 1932.

Simmons, Rick. *Defending South Carolina's Coast: The Civil War from Georgetown to Little River*. Charleston, SC: The History Press, 2009.

Singleton, Theresa A. *The Archaeology of Slavery and Plantation Life*. Orlando, FL: Academic Press, 1985.

Smythe, Augustine Thomas. *The Carolina Low-Country*. New York: Macmillan, 1931.

Stampp, Kenneth M. *The Causes of the Civil War*. New York: Simon & Schuster, 1986.

Talbert, Roy, and Meggan A. Farish. *The Journal of Peter Horry, South Carolinian: Recording the New Republic, 1812–1814*. Columbia: University of South Carolina Press, 2012.

Taylor, Alrutheus Ambush. *The Negro in South Carolina during the Reconstruction*. Washington, D.C.: Association for the Study of Negro Life and History, 1924.

Thomas, Albert Sidney. *A Historical Account of the Protestant Episcopal Church in South Carolina, 1820–1957; Being a Continuation of Dalcho's Account, 1670–1820*. Columbia, SC, 1957.

Thomas, John Peyre. *The History of the South Carolina Military Academy, with Appendixes*. Charleston, SC: Walker, Evans & Cogswell, 1893.

Tindall, George Brown. *South Carolina Negroes, 1877–1900*. Columbia: University of South Carolina Press, 1952.

Tower, R. Lockwood, Warren Ripley and Arthur M. Wilcox. *A Carolinian Goes to War: The Civil War Narrative of Arthur Middleton Manigault, Brigadier General, C.S.A.* Columbia: Published for the Charleston Library Society by the University of South Carolina Press, 1983.

Vernon, Amelia Wallace. *African Americans at Mars Bluff, South Carolina*. Baton Rouge: Louisiana State University Press, 1993.

Warren, Geoffrey. *Fashion Accessories since 1500*. New York: Drama Book, 1987.

Williamson, Joel. *After Slavery: The Negro in South Carolina during Reconstruction, 1861–1877*. Chapel Hill: University of North Carolina Press, 1965.

Winyah Indigo Society. *A Short History of the Winyah Indigo Society of Georgetown, South Carolina, 1755–1950, with Lists of Deceased and Living Members*. Georgetown, SC: Winyah Indigo Society, 1950.

# Index

## K



## L



## M



## N



## O



## P



## R



## S

## T



## V



## W

# About the Author

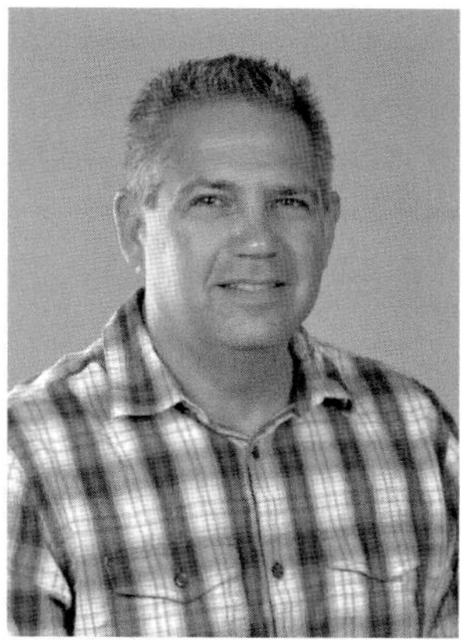

Christopher C. Boyle is a full-time social studies teacher at Socastee High School in Myrtle Beach, South Carolina, and a part-time teaching associate at Coastal Carolina University (CCU) in Conway, South Carolina. He grew up in New York's Hudson Valley and in 1990 began taking classes at Coastal Carolina. Upon graduating from CCU with a bachelor's degree in history, he further studied history at Winthrop University, where he graduated in 1996 with his master of the arts degree in American history. His thesis was entitled "The Social Organizations and Leisurely Activities of the Georgetown Rice Planters: 1840–1861." He lives in Loris and has two children: Brandon and Hannah.